Honey bees,

a natural and a less natural history

by
Jacques van Alphen

Northern Bee Books

Honeybees, a natural and a less natural history

Copyright © Jacques van Alphen

All rights reserved. No part of this publication may be reproduced, stored in a retrieval system, transmitted in any form or by any means electronic, mechanical, including photocopying, recording or otherwise without prior consent of the copyright holders.

Published 2024 by
Northern Bee Books,
Scout Bottom Farm,
Mytholmroyd,
West Yorkshire
HX7 5JS (UK)
Tel: 01422 882751 Fax: 01422 886157
www.northernbeebooks.co.uk

ISBN 978-1-914934-83-4

Design and artwork DM Design and Print

Honeybees,

a natural and a less natural history

by
Jacques van Alphen

Contents

	Page
Introduction Why this book?	1
Chapter 1 From the Palaeogene to the last Ice Age	5
Chapter 2 The annual cycle of the beehive	11
Chapter 3 The division of labour within a hive	15
Chapter 4: A people without leaders	19
Chapter 5 The queen chooses the sex of her children	25
Chapter 6 Dating, where and when the partners meet	27
Chapter 7 Why does a queen mate with so many males?	33
Chapter 8 The paradox of the old queen's departure	37
Chapter 9 The arms race	43
Chapter 10 Why so many drones and so few queens?	47
Chapter 11 The language of bees: shaking and trembling	51
Chapter 12 Moving to a new home, swarming honey bees	57
Chapter 13 Chosing a future home	63
Chapter 14 Honey bees in the wild	67
Chapter 15 Honey bees: indigenous, wild, domesticated or simply kept?	73

Chapter 16 Enemies from the Far East (1): the varroa mite 77

Chapter 17 The evolution of resistance through natural selection 81

Chapter 18 Varroa in South America 87

Chapter 19 Selection for varroa resistance 89

Chapter 20 Enemies from the Far East (2): The Asian hornet 93

Chapter 21 Self-medication in honey bees 97

Chapter 22 Black bees and racism 101

Chapter 23 The honey bee as a competitor to solitary bees 107

Chapter 24 Conservatories 119

Chapter 25 Darwinian beekeeping 123

Chapter 26 Finally 131

INTRODUCTION. WHY THIS BOOK?

A quick search on the Internet for books on honey bees yields more than 100 references. A new book on the subject may therefore seem superfluous. Books on bees are often written by beekeepers who have learned from their peers, in a tradition of passing on the secrets of the trade from generation to generation. This tradition has its roots in the days when beekeeping was a real profession, when bees were still kept in straw skeps rather than wooden hives. In 1874, the British Beekeepers Association was founded, the first organisation to bring together professionals. In 2018 the association counted 26,555 members including 400 professional beekeepers.

There is a tradition of passing on knowledge through courses and conferences, and meetings between beekeepers often give rise to lively exchanges on all aspects of the profession. The danger, however, is that age-old practices are not called into question, or re-evaluated and brought up to date.

A few years ago, when I found myself involved in research to combat an exotic parasitic mite (which could wipe out infected bees if nothing is done to eradicate it), I discovered a world that seemed full of paradoxes. I quickly realised that I would only be able to work properly if I first made up for my lack of knowledge of beekeeping. For almost a year, I read all sorts of publications on the subject, from the oldest to the most recent. I used these readings as a basis for writing a long scientific summary article on the parasitic mite. I attended lectures on bees and started giving them myself at local beekeeping schools, which enabled me to learn and specialise thanks to critical and pertinent questions from experienced beekeepers.

A large part of my professional life has been devoted to behavioural research on parasitoid wasps, which are distant relatives of bees. I tried to understand how evolution through natural selection had shaped their behaviour. At the time, I was already passionately reading all the publications on the evolution of bee behaviour, worthy of the best science fiction novels. For evolutionary biologists, social insects represent a challenge. How do you explain, for example, a female bee that refrains from laying eggs but helps another to raise her offspring? This is why a great deal of research has focused on ants, termites, wasps, stingless bees and honey bees. Of all these social insects, the honey bee is the only one that can be reared in hives with removable combs, allowing all kinds of experimental manipulations. This is why it is the best-studied insect in the world. Thanks to this research, the complex enigma of the social behaviour of bees has largely been solved. It's a subject that is hardly ever covered in books on beekeeping, but it's also a difficult one to understand. The first reason I wrote this book was to make the subject accessible to everyone, especially non-biologists.

The second reason for writing this book is that, as a biologist, I am aware that animal behaviour is adapted to life in the wild and not in the context of human breeding. This applies to pigs and chickens, but also to bees. Little was known about the life of bees, when they were not kept in hives, until Thomas Seeley decided to devote himself to the study of wild honey bees in the United States. His work revealed that honey bees have very different lives depending on whether they are in the wild or kept by beekeepers. The latter use the bees for their own benefit, to the detriment of the insect. A better understanding of the behaviour of bees in the wild could make beekeeping healthier and more respectful of bees.

In North America, honey bees did not occur naturally. The wild bees that Seeley studied were therefore hybrids of many different subspecies, imported from various parts of Europe, Asia and Africa. In Europe, the honey bee is a native species, with several subspecies that do not normally interbreed because they are separated by mountain ranges. They are adapted to the local flora and climate, and cohabit with other nectar-feeding and pollinating insects. Unfortunately, beekeeping in Western Europe, with its intensive use of hybrid bees and imported subspecies, is threatening the native black honey bee with extinction. The third reason that led me to write this book is a plea for the preservation of the black bee.

A fourth reason for this book is the fight against the exotic parasitic mite varroa. For over twenty years, the solution has existed and, slowly but surely, is being adapted and applied: honey bees are becoming more resistant. The initial lack of natural resistance to varroa raises the question of beekeeping practices: bees are selected to be not aggressive and to produce a lot of honey. This is why beekeepers have largely deactivated natural selection against infectious diseases. So, it is time to think about improving practices.

As a biologist and an outsider, I was amazed at what beekeepers take for granted. For example, that in the Netherlands, three of the Wadden Islands, which have National Park or Natura 2000 status, are authorised to be used for the exclusive cultivation of exotic bees. Or the fact that beekeepers consider it normal to release their bastard or exotic drones into the countryside, thus eliminating the native bee through genetic pollution. Or, more importantly, that all the new knowledge about bee behaviour, genetics, ecology and evolution has not led to fundamental changes in beekeeping.

A final reason for writing this book is that, in recent years, beekeeping has often received a bad press. Beekeepers have been accused of being a threat to the biodiversity of other pollinators, particularly solitary bees. Can these accusations be heeded, or do they conceal a wider and more complex problem? This is what I intend to investigate.

The knowledge gained from bee research provides a good starting point for new practices that will give bees greater resistance to disease and parasites, enable them to behave more naturally and encourage beekeeping that is more respectful of biodiversity. The final chapter of this book will allow me to set out my – utopian (?) view of the future of European beekeeping.

CHAPTER 1:
From the Palaeogene to the end of the last Ice Age

Looking a little further back than our brief history, we can see that we are living in a period of glaciation that has lasted for some 2.4 billion years. Periodically, large ice caps form over Eurasia and North America. These periods of glaciation always last around 100,000 years, interspersed with relatively short, warm interglacial periods. We are most familiar with the last complete cycle of the ice age, which extends from 135,000 years ago to the present day, and in particular with the last warming period, which began around 16,000 years ago and is responsible for the current warm interglacial climate. Researchers estimate that temperatures initially changed rapidly, by 10 to 12% in five to ten years. Such major and rapid climatic changes also alter the distribution pattern of species. Research on this subject shows that there is a close relationship between climate fluctuations and changes in vegetation.

The major mountain ranges of central and southern Europe generally extend from east to west. This is the case of the Cantabrian Mountains, the Pyrenees, the Alps, the Transylvanian Alps and the Caucasus. These mountain ranges all had extensive ice caps during the last ice age. Further south, also running east-west, are the Mediterranean and the Black Sea, with the rather mountainous peninsulas of Iberia, Italy, Greece and Turkey. From these southern mountain ranges, a plain of permafrost, tundra and cold steppe stretched to the Urals into eastern Russia. Even today, the seas and mountains represent formidable barriers for most organisms, and the mountain ranges, when covered in ice during the last ice age must have been even more difficult to cross. Plant and animal remains show that at the height of the Ice Age, 18,000 years ago, most of the organisms present in Europe today had retreated further south, towards the Iberian Peninsula, Italy and the Balkans, and perhaps even the Caucasus and the Caspian Sea for some. Around 16,000 years ago, the climate began to warm, the ice began to retreat and species (including probably the honey bee) began to expand their habitats from south to north, and around 13,000 years ago, plant and tree species began to appear further north. This rapid advance was slowed for 1,000 years, around 11,000 years ago, by a short period of cold followed by a warming of the climate, which allowed vegetation to expand rapidly northwards around 10,000 years ago. Around 6,000 years ago, the structure of the vegetation was already very similar to that of today, and it is likely that the honeybee settled again in our regions at that time.

The evolution of honey bees is also a long story. The oldest data date back 66 million years to the Palaeogene, when the first fossil bees were discovered. A comparison of the wing veins of these bees with those of today has shown that the first ancestors of honey bees came from Europe and from there colonised Asia, Africa and North America. They then became extinct in North America and Europe, but it was in Asia, between 12 and 10 million years ago, that the species that exist today evolved.

Eleven species of honey bee live in Asia today, including our own species, the European honey bee, *Apis mellifera*. The common ancestor of *A.cerana* and *A. mellifera* originated from Asia, *A.mellifera* split off from its ancestors six to eight million years ago and evolved in Africa. It is the only species found today outside Asia, in Africa and Europe. By comparing the DNA of these different honeybee populations, it is possible to reconstruct the link between them and understand how they spread from Asia.

On the basis of shape and DNA characteristics, some 29 subspecies of European bees have been identified. They are divided into four groups, which differ in their distribution and physiological, ecological and behavioural characteristics. These groups from Africa (A), Western Europe (M), Eastern Europe (C) and Asia (O) separated between 0.7 and 1.3 million years ago. DNA data allow us to reconstruct their routes to Europe. The data suggest that the M lineage reached Europe from Africa via the Iberian Peninsula, and that the south-eastern European subspecies of the C lineage arrived from Africa via the Middle East, Turkey and the Balkans.

The Western European black bee, along with the Spanish subspecies, belongs to group M, while the Italian subspecies *A.m. ligustica* and the Eastern European subspecies *A.m. carnica* belong to group C. The different groups of the European honeybee appeared when Western Europe was under the influence of successive ice ages interspersed with warmer periods. During the ice ages, group M bees retreated south of the Pyrenees, while lineage C bees moved behind the Alps into Italy and the Balkans.

The mountain ranges behind which the bees retreated during the ice ages have always represented major obstacles to the spread of the different subspecies. Irati Miguel and his colleagues have shown that after the last ice age, the black bee was only able to cross the Pyrenees at their western and eastern ends, where the mountains are lower. The Pyrenees therefore still represent a barrier between the black bee *A.m. mellifera* and the Spanish subspecies *A.m. iberiensis*. In the same way, the black bee is separated by the Alps from the Italian subspecies *A.m. ligustica* and by the Transylvanian Alps from the Eastern European subspecies *A.m. carnica*. By their very nature, these exotic subspecies would never have reached Western Europe. But the introduction of exotic bees by beekeepers are putting the western black bee in serious danger of extinction, due to continuous mating with these exotic bees.

The area originally populated exclusively by the black bee extends from the Pyrenees to southern Scandinavia and from the British Isles to the Urals. In this immense geographical area, the climate varies from Atlantic to Mediterranean and continental. Black bee populations in different parts of the range appear to have adapted to the local climate, flora and other pollinators. The movement of bee colonies between these areas disrupts this local adaptation. Wall paintings dating back 8,000 years found in a Spanish cave show that man was already exploiting wild bees at that time. In Western Europe, bees were kept in hives from the early Middle Ages. This long period of exploitation probably had little effect on the genetic characteristics of local honeybee populations. In fact, people kept local bees, which could mate freely with their wild counterparts and were part of the same population. The population structure of black bees with other locally adapted bees thus remained intact until the industrial revolution. Watt's invention of the steam engine was followed by the introduction of steam-powered trains and ships in the 19th century, which facilitated long-distance transport. This led to an international trade in bee colonies, which slowly but surely led to the destruction of virtually all local black bee populations, except the Irish.

Recommended reading

Goffrey M Hewitt, 1999. Post-glacial re-colonization of European biota. Biological Journal of the Linnean Society 68: 87-112.

Fan Han, Andreas Wallberg & Matthew T. Webster, 2012 From where did the Western honeybee (Apis mellifera) originate? Ecology and Evolution 2012; 2(8): 1949–1957

Charles W. Whitfield, Susanta K. Behura, Stewart H. Berlocher, Andrew G. Clark, Spencer Johnston, Walter S. Sheppard, Deborah R. Smith, Andrew V. Suarez, Daniel Weaver, Neil D. Tsutsui. 2006. Thrice out of Africa: Ancient and Recent Expansions of the Honey Bee, *Apis mellifera*. Science 314: 642-645.

Irati Miguel, Mikel Iriondo, Lionel Garnery, Walter S. Sheppard, Andone Estonba, 2007. Gene flow within the M evolutionary lineage of Apis mellifera: role of the Pyrenees, isolation by distance and post-glacial re-colonization routes in the western Europe. Apidologie 38 (2007) –155

The distribution of different subspecies of *Apis mellifera*

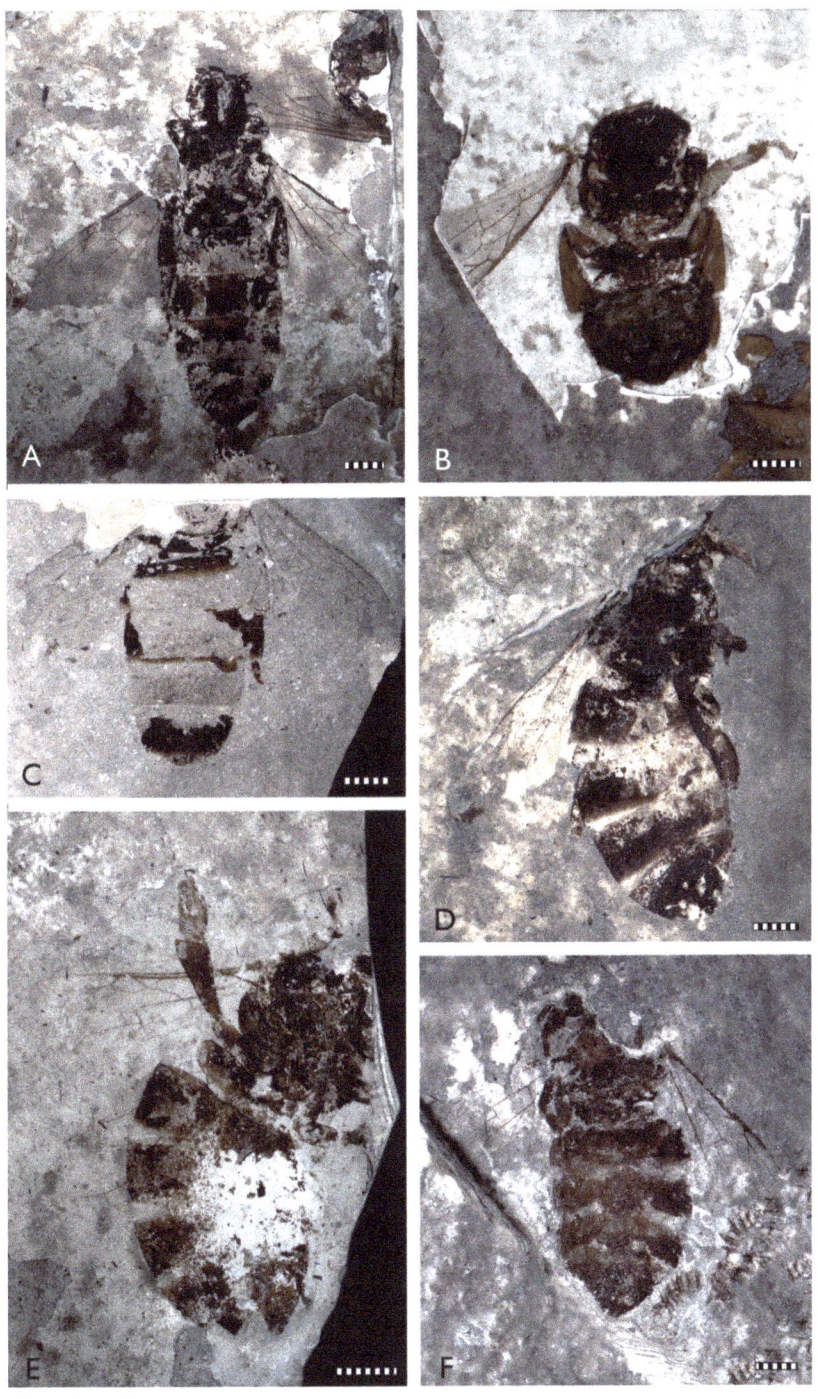

Apis armbrusteri, a honeybee from the Miocene found in Randeck, Germany.
ZooKeys 96: 11–37 (2011) doi: 10.3897/zookeys.96.752

CHAPTER 2.
The annual cycle of the hive

Reproduction is the big business in nature, with everyone wanting to ensure their offspring have a place in the next generation. A bee colony's contribution to this objective is the production of a certain number of swarms, plus a large number of male bees, the drones, of which some will mate successfully with young queens from other colonies. Unlike bumblebees and wasps, a honeybee colony does not die after reproduction: they can hibernate and reproduce again. To survive the winter, a colony needs to grow again after swarming, so that it has enough workers. This is necessary because honey bees do not hibernate like most insects in our climate, but they fight the cold by burning honey. To minimise the cost of heating, they huddle together to form a ball. If this "winter cluster" is too small, they are unable to keep warm. As bees also die during the winter, and it is very important to have a sufficient number of workers when the nectar- and pollen-producing plants flower in spring, the bees start to reproduce well in advance of this flowering. This means that, from January onwards, they raise the temperature of the winter swarm to 35°C, so that the eggs and larvae can develop properly. This operation entails additional heating costs. A hive of around fifteen thousand bees burns around twenty kilos of honey during the winter. If the winter supply is too low, the hive dies of starvation and cold without being able to replenish itself in the spring.

    A successful season begins in early spring with the collection of nectar and a large quantity of protein-rich pollen, which is used to rapidly produce a number of young workers, followed by drones and young queens. When all the flowers bloom a little later, these new workers can already be employed to reap the abundant harvest. Breeding colonies that are strong enough to swarm do so at the end of April or the beginning of May, so that the swarmed bees and those remaining in the hive can use the rest of the season to build their colony and build up sufficient reserves for the winter.

    These bees need to gather food over a much longer period than solitary bees, which have to lay eggs in a short space of time and look after the resulting larvae until they begin to pupate. They therefore live on a very different spatio-temporal scale to solitary bees. They seek out the areas richest in nectar and pollen for a quick harvest, and ignore less productive areas. By placing the hives on a scale, it is possible to see how the nectar and pollen harvest varies over time. In some areas, there is a rapid increase in the weight of the hive from the start of the spring bloom. As honey bees don't gather pollen and nectar in bad weather, we can see a loss of weight at this time. The bees then draw on their reserves. But when the weather is fine (i.e. if the temperature is above 10°C and it's not raining), they make foraging trips. In spring, nectar- and pollen-producing flowers are plentiful almost everywhere, but in some places the supply diminishes considerably once the trees and bushes have finished flowering at the end of June. Honey bees get round this problem by ensuring they have enough reserves to survive until the following spring. They also reproduce less, to reduce the number of brood they have to feed.

On heaths and salt marshes, there is a profusion of nectar-producing flowers in late summer. This is the time when heather and sea lavender bloom. Jean Louveaux compared black bee populations in the Landes department in France, which, as its name suggests, is largely covered by heathland, with populations in two areas without heathland, near Paris and Avignon. Around these towns, the flowering of nectar-producing and pollinating plants is concentrated in spring, and there is little left to harvest after the chestnut trees flower at the end of June. While colonies in the Paris and Avignon regions reduced brood production in June, those in the Landes region continued to produce new workers until the end of August. Louveaux then experimented with moving colonies from one area to another. His observations showed that the high production of worker bees at the end of the season was characteristic of the Landes bees. This was not a reaction by the bees to the arrival of heather nectar, as they had anticipated the nectar harvest several weeks in advance. It takes three weeks for a worker to develop, and she remains in the nest for more than two weeks before coming out to look for food. The bees must therefore start raising additional workers at least five weeks before the heather flowers. Jean Louveaux, who also monitored the weight of the hives, was able to demonstrate that bees from the Landes region were adapted to the late flowering of the heather and harvested much more honey than those from Paris and Avignon. Conversely, the latter were more successful in their own region. Native black bees are therefore adapted to the flowering periods of local nectar-producing plants. Experiments also shows that it doesn't matter when winter reserves are built up, as long as they are sufficient at the end of the flowering season. Most insects, including solitary bees and bumblebees, spend the winter in a state of inactivity where their body temperature matches that of the environment. Special physiological adaptations prevent them from dying in conditions of intense cold. The advantage of this form of hibernation is that it requires very little energy. Honey bees, probably because of their tropical origins, have found an energy-intensive answer to the problem of winter survival. Although they prefer to choose well-insulated nesting sites, and the formation of a "winter cluster" is an attempt to minimise energy losses, a third of their annual energy budget is spent in winter on heating costs. In addition, the hibernation mode of honey bees allows them to harvest a lot of food very early in the season. Bumblebees have a slower start, even though the young queens fly early and at lower temperatures than honey bees. By the time they have found a nesting place and reared their first workers, the willows have already finished flowering. Bumblebee colonies therefore do not number in the thousands.

Recommended reading

J. Louveaux, M. Albisetti, M. Delangue, M. Theurkauff, !966. Les Modalités de l'adaptation des abeilles (*Apis mellifica* L.) au milieu naturel. Les Annales de l'Abeille, INRA Editions, 1966, 9 (4), pp.323-350. hal-00890241.

CHAPTER 3.
Distribution of tasks within a hive

The Ghent writer Maurice Maeterlinck (1862-1949) was a passionate beekeeper, so fascinated by the enigmas of social insects that he published a book on the lives of bees (in 1901), then termites (in 1926) and finally ants (in 1930). It seems that he plagiarised the South African Eugène Marais to write part of his book on termites - disappointing behaviour that you would not expect from the 1911 winner of the Nobel Prize for Literature!

Maurice Maeterlinck described the life of a colony of bees as if they were humans. He wrote about the extraordinary discipline of the bees, the self-sacrifice of the workers, their dedication, their affection for their queen, their faith and hope that never wavers. He couldn't really understand the complexity of a bee colony and spoke of the "frightening complexity of the most natural phenomena" and the "incomprehensible organisation". He also noted that a population of bees could carry out complex tasks and therefore thought that there must be a force in charge of organisation. He called this "the spirit of the hive". Surprisingly, a few years later, Eugène Marais entitled his book on termites (published in 1926) "Die Siel van die Mier" ("The Soul of the Ant"). He too suspected the presence of an organising force in the functioning of a termite colony.

Years later, the biologist and beekeeping researcher Robert Page published a book whose title, "The Spirit of the hive: the mechanisms of Social Evolution", was borrowed from Maurice Maeterlinck. In this book, he summarises research into the incredible behaviour of bees. It provides a complete picture of the genetic, physiological and behavioural mechanisms underlying the division of labour in bee colonies and their complex social behaviour. The conclusion is that there is no organising force, so the spirit has been eliminated.

The first level of division of labour is the three castes. A larva will become a drone, a queen or a worker depending on three factors: the fertilisation of the egg, where it is laid (fertilised or not), and the type of food the growing larva eats.

The queen lays unfertilised eggs in the large cells of the drones, and fertilised eggs in the smaller cells where the workers are growing. She also lays fertilised eggs in special bowl-shaped cells with a downward-facing opening, which the workers build at the base of the combs in spring. The queens develop in these cells, which the workers enlarge into elongated cylinders. At first, the workers feed the three types of larvae with a secretion rich in proteins and sugar from the maxillary glands, called royal jelly. When they are older, the larvae destined to become workers or drones are fed mainly nectar and pollen, their diet being lower in sugars and proteins. Young queens are fed royal jelly until they reach adulthood. As a result, they develop much more rapidly and become larger than the workers. Young drones develop more slowly, like workers, but grow longer and become larger than workers. The rapid development of queens (twelve days) is all the more important because, if the queen of a colony dies or is replaced, the colony will no longer function properly. A healthy queen can lay up to 2,000 eggs a day: every day without a queen limits the growth of a colony and therefore its chances of survival the following winter.

We might therefore ask why the nurse bees don't continue to feed the young workers with royal jelly, so that they too can develop more quickly and enable the colony to grow more rapidly. Switching from royal jelly to a diet lower in sugars and proteins is apparently a compromise. If all the larvae ate only royal jelly, many more workers would be needed to produce it - workers who are in short supply in early spring, when the population needs to grow rapidly. It is therefore preferable to rear a larger number of workers simultaneously, even if it takes longer.

The work of the drones is purely reproductive; they try to mate with young queens from other colonies. The queen's job is to make one or more nuptial flights to mate and reproduce by laying eggs.

All other tasks in a colony are the responsibility of the workers. The distribution of these tasks among them, according to their age, largely determines how the colony functions. A young worker first works in the hive close to her place of birth: she begins by cleaning the cells in which the worker bees have grown. After three days, when her maxillary glands begin to secrete energy-rich food for the young larvae and the queen, her career as a nurse bee begins. Her job then consists of feeding the larvae, the other workers and looking after the queen, as well as sealing the cells in which the larvae transform into chrysalids and cleaning those of the other workers. She does this until she is around twelve days old. She then leaves the brood nest for the edges of the combs, where the honey is stored. Here, near the entrance to the hive, she removes nectar from the returning workers, transforms it into honey and stores it in the available cells . She also collects the pollen and helps to ventilate and guard the entrance to the nest. If the nest needs to be enlarged or repaired, she can also produce wax. When she is around twenty days old, her work outside the nest continues until she dies: searching for and collecting nectar, pollen, water and resin, and sometimes finding a new place to live (see chapter 12). The age criterion in the division of labour is just one of the ways in which a population of bees can function as a super-organism. This division of labour is not rigid, and can vary according to the needs of the moment. Workers spend part of their time walking or being idle, and can be recruited when needed. When nectar and pollen are scarce, the bees start working outside the nest later. On the other hand, if a new and important source of nectar and pollen is detected by the scouts, workers can be recruited to collect it. If attacked by hornets, the workers, alerted by the guards, rush to the nest entrance to defend it.

Genetic differences between workers also play a role in the division of labour. Some workers specialise in certain tasks, such as collecting pollen or water, while others guard the nest entrance. This genetic ability to specialise probably makes the workers more efficient in their tasks. This is why, in order to multiply specialisation, it is important for a colony to have many different fathers. These colonies then seem to function better (see chapter 7).

Recommended reading

Page RE, 2013. "The Spirit of the Hive: The Mechanisms of Social Evolution. Harvard University Press.

Maurice Maeterlinck, 1901. La vie des abeilles. Paris, Bibliothèque-Charpentier, Eugène Fasquelle Editeur.

Thomas D. Seeley, 1985. Honeybee Ecology. A Study of Adaptation in social Life. Princeton University Press, Princeton, New Jersey.

Thomas D. Seeley, 1995. The Wisdom of the hive. Harvard University Press, Cambridge Massachusetts.

CHAPTER 4:
A hive without leaders

In his History of Animals, Aristotle wrote that bee populations are ruled by rulers. He thought that the rulers might be females, but he wasn't sure, nor how many there were. *"These rulers have the abdomen, or the part behind the waist, shorter, and they are called 'mothers' by those who think that they bear or give birth to the bees... Others, again, claim that these insects copulate, and that the males are the drones and the females the bees."* This quotation clearly refers to the sex of drones and workers, but the next quotation concerns queens: *"The common bee is formed in the cells of the comb, but the queen cells, six or seven in number, are situated at the bottom of the comb, hanging distinctly from each other, in which the larvae develop in a very different way from the common brood."* A little further on in the same text, Aristotle attributes sovereignty to the queens: *"All bees have a stinger, except the drones. The leaders have a stinger, but never use it; this explains why some believe they have no stinger"*. Although Aristotle himself was unsure of the sex of the rulers, and others after him suggested that the ruler was a female, it was long accepted that the leader of a colony of bees was a male, *i.e.* a king. It was only at the end of the 17th century, when Jan Swammerdam's[1] detailed drawings of the reproductive organs of a queen bee were published (posthumously) by Boerhave, that it was accepted that this was a female bee, and therefore a queen. The Aristotelian view that the queen is the ruler of a population of bees can still be encountered today.

The idea of "leadership" would imply a conflict of interests between the queen and the workers, with the former prevailing by suppressing the latter. Is this hypothesis correct? Queens constantly produce a pheromone from the glands in their jaws. This pheromone has several functions, one of which is often mentioned: "It stops the development of the ovaries of the workers". The queen thus has a serious weapon at her disposal to prevent the workers from laying eggs themselves. In colonies without a queen, the ovaries of the workers develop and they can then lay eggs. But this does not mean that there is a conflict of interest between the workers and the queen when the latter is present. Such a conflict would exist if the workers would be better off laying eggs and raising their own offspring, instead of helping the queen to raise her children, i.e. the brothers, sisters and half-sisters of the same workers. It's not simple to find out if that is so.

Reproduction is always linked to the number of copies of one's own genes that are passed on to the next generation. A bee is always more closely related to its own children than to its brothers and sisters, so the proportion of its own genes passed on to a brother or sister is lower than to its own child. This could only be compensated for if the number of siblings raised by the workers helping the queen was greater than twice the number of children the worker could have herself. Thomas Seeley gives some arguments in favour of this scenario. One of these arguments is that the fertility of the queen is much higher than that of a worker. However, this remains in the realm of speculation, as female workers do not generally lay eggs or mate, so this hypothesis cannot be seriously studied. We can, however, provide an answer because, in the very south of Africa, there is a subspecies of bee whose workers can have daughters. In the Cape honey bee, workers can lay eggs without ever having mated, with two sets of chromosomes. The daughters of these workers are genetically identical to their mothers. These bees lay eggs themselves when the old queen leaves with a swarm, leaving the reproductive function to the queen for the rest of the year. When the old queen swarms, the bees lay eggs in both worker and queen cells, enabling their daughters to become young queens. If they didn't do this, the new queen would be a daughter of the old queen. The relationship between the workers and a daughter of the old queen is only half that of their relationship with the old queen herself. In this situation, it is preferable to reproduce yourself. If a worker's daughter were to become the new queen, she would make a nuptial flight and mate with a large number of drones. As a result, the workers she produced would be only half related to the worker that was the mother of the queen.

The interesting thing for us is, that the workers who can have daughters still like to leave reproduction to the queen for the rest of the year. That behaviour can only arise and persist if there is no conflict over this division of labour. Therefore, we conclude that the queen and the workers have no conflict of interest over who lays the eggs. The idea of the queen as the leader of a population of bees also suggests that she somehow directs the activities of the hive. However, this is not in line with reality either. The queen's role is to lay eggs. She is not involved in managing the flow of information about where to find food and water, or in allocating workers to different tasks. She doesn't even have access to important decisions such as swarming and moving. To say that the queen suppresses the development of worker ovaries is not a correct interpretation. By producing pheromones with her maxillary glands, the queen informs the workers that she is present and continues to lay eggs. A second argument against this interpretation is that, if the queen is removed from a colony but larvae remain in the combs, the workers, instead of starting to lay eggs, will rear some of the young larvae to become new queens. The ovaries of the workers do not develop in the presence of brood. It is only in the absence of brood that the workers begin to lay eggs themselves. In this case, the bee colony is "desperately queenless" and in danger of disappearing. The only thing to do is to produce drones from worker eggs. With a bit of luck, these will mate with young queens and the colony's genes will still be passed on. However, there are conflicts of interest when it comes to reproduction between workers with different fathers. It is not advisable to rear the son of a worker from another father, because the genetic relationship with this son is very limited. Instead of looking after this type of egg and its hatching larva, the workers will generally remove the eggs of other workers. This makes it very difficult for them to reproduce. In a way, they police each other.

Sometimes, genetic differences between workers lead to a loss of efficiency, as there is a form of struggle between them. For example, only certain workers are capable of opening cells containing diseased and infected pupae and removing them. This is because they have the genetic disposition to perform this useful work. If a worker has not inherited this ability from her father, she will quickly close a cell that has just been opened in order to extract the pupa. In doing so, she runs counter to the hygienic behaviour of other bees (see chapter 19).

However, a population of bees is perfectly capable of functioning without a manager. Each worker does her job, depending on where she is, her age and by responding to information about the situation inside and outside the hive. And the queen lays eggs.

[1] Jan Swammerdam (1637-1680) was the son of an Amsterdam apothecary. He had a large naturalist collection, which sparked his son's interest in nature and science. Swammerdam's father intended his son to become a priest, but nevertheless agreed to let him study medicine. During his studies, Jan Swammerdam discovered a passion for anatomical research and learned to use microscopes. He used those of Johan Jooster van Musschenbroek (1660-1707), a manufacturer of optical instruments in Leiden and father of the physicist Pieter van Musschenbroek. After completing his studies, Jan Swammerdam never practised as a doctor, but lived at his father's expense and continued his research into insects.

Recommended reading

Thomas D. Seeley, 1995. The Wisdom of the hive. Harvard University Press, Cambridge Massachusetts.

Madeleine Beekman, Michael H. Allsopp, Lyndon A. Jordan, Julianne Lim and Benjamin P. Oldroyd, 2009. A quantitative study of worker reproduction in queenright colonies of the Cape honey bee, *Apis mellifera capensis*. Molecular Ecology (2009) 18, 2722–2727

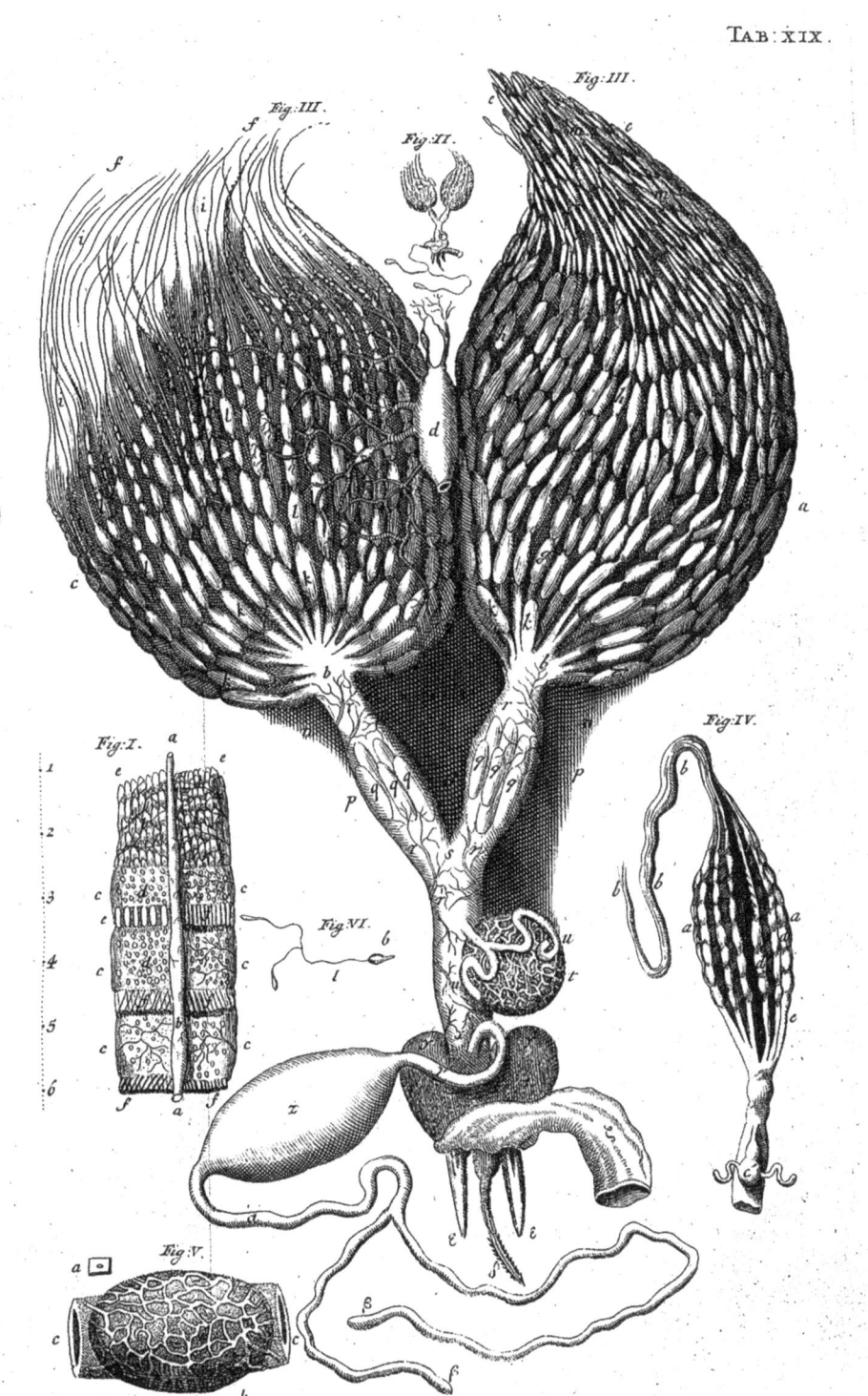

Swammerdam's drawing of the queen bee's reproductive organs, as observed through the microscope

CHAPTER 5.
The queen chooses the sex of her children

One of the demands of the feminist movement in the 1960s was the right to abortion. The slogan was: "Women decide". It was, and still is, the wish of certain future parents not only to have a child, but above all to have a son or a daughter, and therefore to be certain of the sex of the unborn child from the moment of conception. This is impossible for humans. Except in the controversial case of in vitro fertilisation followed by the selection of an embryo of the desired sex. But this method is banned in the UK and many other countries. The impossibility of choosing the sex of the future child results from the way in which the sex of an individual is determined genetically. In humans and mammals, the genes that determine sex are located on specific chromosomes, the X and Y chromosomes. An embryo with one X and one Y chromosome will generally be male, while an embryo with two X chromosomes will be female. In birds, the reverse is true: males have two identical chromosomes while females have two different ones. In insects, sex inheritance can vary.

In bees and all other hymenoptera (such as bumblebees and wasps), sex is determined as follows. After mating, the eggs are not immediately fertilised, but the future mother stores the sperm in a sac-like organ (spermatheca). When she lays an egg, she can decide whether or not to open the sperm sac and thus allow a spermatozoon to reach the egg. An egg fertilised by a spermatozoon has two sets of chromosomes, one from the mother and one from the father. It generally develops into a female. What is special is that eggs that are not fertilised and have only one set of chromosomes, develop into males. The mother can therefore decide the sex of the child. In honey bees, the mother is the queen. The workers in a bee colony are her daughters and the drones her sons. The young queens, who have been given different food from the workers, are also her daughters. Workers and queens therefore have two sets of chromosomes, one from the queen and the other from the drone that provided the sperm. Drones have received only one set of chromosomes from the queen, and are therefore true children of their mother.

This method of sex determination gives the queen the freedom to choose whether the egg she lays will be male or female. But this freedom also has a limit.

Honey bees and certain hymenoptera have a special and complex system in common: a single gene determines sex, and for an embryo to become female, it must have two different variants of that gene. If it has only one variant, it becomes male. This is why an unfertilised egg always becomes male. If the embryo receives the same variant twice from its mother and father, it will become a non-functional male, with two sets of chromosomes instead of one. This type of male is quickly eliminated by the workers. This is one of the reasons why bees must avoid inbreeding, which risks increasing the proportion of "failed" males. This explains why the queen mates with a large number of different drones and can make several mating flights (see chapter 6 & 7). This is how, unlike humans and many other animals, bees, ants and wasps can decide the sex of their offspring, by fertilising or not fertilising an egg!

While the queen decides for each egg she lays whether it will be fertilised or not, and therefore whether it will become male or female, it is the workers who indicate which type of egg should be laid in an empty cell: fertilised eggs in the small worker cells and suspended queen cells, unfertilised eggs in the large drone cells. Together, they determine the sex ratio (see chapter 10).

Recommended reading

R.H. Crozier, 1971. Heterozygosity and Sex Determination in Haplo-Diploidy
The American Naturalist 105: 399-412. DOI.org/10.1086/2827

CHAPTER 6.
Dating: where and when finding partners?

On 28 June 1792, the Reverend Gilbert White (1720-1793) walked from his rectory in Selborne, England, to his brother's house. When he reached the top of the hill, he heard a loud buzzing of bees, but did not see a single one. The sound was different from that produced by a swarm, but he couldn't explain its origin. He did, however, realise that he had just made an interesting observation. Reverend White was the author of "The Natural History and Antiquities of Selborne" and well known for his keen sense of observation and accurate descriptions of natural phenomena. So, he wrote down this observation as follows:

> *"Buzzing in the air. This is a natural phenomenon that can be observed at the top of the hill on hot summer days, which always amuses me, without my being able to explain it satisfactorily. It is an audible buzzing of bees, although not a single insect can be seen. This sound is clearly heard in the vicinity, from the Moneydells to the gate of the lane leading to Mr White's. You'd think there was a large swarm of bees on the move, right over our heads. This sound was heard last week, on 28 June to be exact."*

It took 160 years to find an explanation for this buzzing sound. Beekeepers had observed that on fine spring and early summer days, the drones always took to the air in the early afternoon. Cyprien Zmarlicki and Roger Morse wanted to know where they went next. Among beekeepers, legends were circulating about the gathering places of the drones. If true, this could mean that virgin queens would go to these haunts to mate. The possibility of observing the behaviour of drones and queens in these places could unlock the secrets of mating flight. Some beekeepers thought that the virgin queen mated in the air, close to her nest, while others put forward the hypothesis that drones were prowling around in the hope of meeting a virgin queen. Researchers enclosed a virgin queen in a cage attached to a helium-filled balloon, releasing it in different locations. In general, nothing happened, but in places, many drones approached the cage - and the queen. This proved that certain places are gathering places for drones.

In the spring of 1970, Beowulf Cooper, accompanied by other beekeepers, went to Selbourne and took the same walk as Reverend White, this time armed with a helium balloon to which was attached a cage with a virgin queen inside. At the exact spot where Reverend White had heard the buzzing, a large number of drones rushed to the cage. Stephen Fleming repeated the experiment on 25 June 1977, 225 years after Gilbert White's observation, and again observed a gathering of drones. Selbourne is therefore very probably the oldest known site for this phenomenon.

However, this answer to Reverend White's question raises other questions. The first is to understand how the drones and young queens know the location of this meeting site. Drones only live for a few weeks, and a queen only makes a short mating flight at the beginning of her life. They have neither the time nor the opportunity to pass on information about these meeting places to the next season's drones and young queens. Nevertheless, the event continued year after year. The brothers Hans and Friedrich Ruttner set out to explain this phenomenon. They studied the mating behaviour of honey bees in a narrow valley in Austria. After marking the drones with a colour specific to each colony, they captured some at the gathering sites they knew. This is how they discovered that the drones flew towards the lowest point on the horizon. This is useful in the mountains, as the flight is directed towards the passes, but it does not explain the gatherings of drones in certain places. Moreover, the usefulness of this behaviour is only obvious in the mountains, and not on the plains, where this method of orientation cannot be used. The drones must therefore have another method.

Gerald Loper and his colleagues decided to study this problem in a different way, using radar tracking of drones flying to and from a gathering place. They found that the drones used elongated features of the landscape as landmarks, flying parallel to roads or rows of trees. At the intersection of these lines, there were often staging areas. But there's more to it than that. As far back as 1902, Slaten thought that drones, like the queen, produced scented substances to attract each other. Today, biologists call these odorous substances "aggregation pheromones". This explains why dancing mosquitoes manage to stay together in a compact cloud. For a long time, there was insufficient evidence to support Slaten's suspicions about these gatherings, until the discovery of these aggregation pheromones. The drones are attracted by the smell of a group of their fellow drones, as well as by the smell of extracts from the maxillary glands, and even by the main components produced by these glands. All these observations prove that the odoriferous substances released by drones attract others and encourage them to stay at the gathering site. This also seems to work to attract virgin queens and lead them to the meeting place.

Once the problem of the meeting place had been solved, the question remained as to the geographical origin of the participants. The Ruttner brothers, having marked the drones with a colour according to their nest of origin, found that they came from different hives, sometimes very far away. Drones can be very numerous on a dating site: Emmanuelle Baudry estimates that there were up to 11,000 on the day she counted them. She also estimated that they came from around 240 different hives. The distance travelled by the drones to the staging site is often between 4 and 5 kilometres, but can be as much as 11 kilometres. As the queens also have to fly between 2 and 4 kilometres to the meeting site, it can happen that a queen mates with a drone that comes from a distance of 15 kilometres from her nest. If the average distance between the nest of a virgin queen and that of a drone is 7 kilometres, then drones are recruited from a zone of 150 square kilometres!

Now that we know the play ground (the place where the drones gather) and the players (the thousands of drones gathered from far and wide, and the occasional virgin queen), it remains to describe the game and its rules. The first game is played on a warm, sunny day with a gentle breeze. Players stay at home when the weather is not suitable. The drones leave their nests at around 1pm, and the queens who want to make a nuptial flight start at around 2pm. By this time, the drones are already circling the meeting place. With their big eyes, they spot anything that comes near. If you use a catapult to throw a small stone into the cloud of drones (as Thomas Seeley did), you can see that they will try to follow and overtake the stone. The reaction of the drones to the arrival of a queen is spectacular. Queens produce sexual pheromones in their maxillary glands, especially when they are virgins. As the virgin queen flies at high speed through a cloud of drones, she leaves behind a cloud of sex pheromones: the drones react quickly and immediately set off on the hunt. The queen then looks like a comet with a tail of drones chasing her. Drones have a bendable sex organ called the endophallus. Once the fastest drone has grabbed the queen, it immediately inserts its sex organ into the queen's wide-open abdomen. The endophallus, containing the spermatozoa, then ruptures and the drone falls backwards, dying. The queen continues her nuptial flight, mating with up to twenty different partners before returning to her colony half an hour later. She stores the sperm from the various drones she has mated with in the spermatheca. This stock is sufficient for her entire life.

The question remains: why has natural selection led to such a mating ritual, in which thousands of drones from hundreds of bee colonies in an area of over a hundred square kilometres come together to compete for mates with rare queens? If each of the 240 colonies, spread over an area of one hundred and fifty square kilometres, produced two young queens, that would leave just 480 mating opportunities for an entire season. If the drones had to search individually for queens, they would have to travel enormous distances to cross this area, with very little chance of success. It therefore seems more sensible for the drones to group together and attract the queens to a single location. The queens, mating with a very large number of drones, would need a lot of time to find a sufficient number of partners. Another advantage of this ritual is that the queens can be selective in their choice of drones, given the simultaneous presence of a large number of candidates. Finally, flying outdoors for a long period of time can be dangerous for the queens and drones, because of predators (hungry swallows and hunting dragonflies) looking for prey.

As queens recruit their sexual partners from a vast territory and mate with many males, genetic traits are exchanged over a long distance and mixed at each generation, rather like a pack of cards that is shuffled before each draw. As the areas where drones and queens congregate overlap, the exchange of genetic traits and the subsequent mixing takes place over a much wider area. In this way, honey bees avoid inbreeding to a considerable extent. The reason for this is explained in chapter 9. It also explains why virgin queens do not mate close to their own nest.

The final question is why natural selection has led to queens that can sometimes mate with up to twenty partners. The most obvious answer, that a single drone would not provide her with a life-long supply of sperm, is not the right one: a drone can transmit up to eleven million spermatozoa during mating, whereas the queen only stores around five million after mating with around twenty drones. Paternity research shows that the sperm stored by the queen comes from all her partners. It is apparently important for the queen's daughters to come from several different fathers (see chapters 7 and 9).

The Reverend Gilbert White would certainly have appreciated knowing exactly what happened over his head on 28 June 1792, and that his observation of the humming in the air would later lead to important discoveries.

Recommended reading

Hans Ruttner and Friedrich Ruttner, 1972. Untersuchungen *über* die Flugactivität und das Paarungsverhalten der Drohnen. Apidologie 3: 203 -232.

Gerald M. Loper, Wayne W. Wolf and Orley R. Taylor, Jr. 1992. Honey Bee Drone Flyways and Congregation Areas: Radar Observations. Journal of the Kansas Entomological Society 65, No. 3: 223-230

Emmanuelle Baudry, Michel Solignac, Lionel Garnery, Michael Gries, Jean Marie Cornuet and Nikolaus Koeniger 1998. Relatedness among Honey bees (*Apis mellifera*) of a Drone Congregation. Proceedings: Biological Sciences 265: 2009-2014

Annette B Jensen, Kellie A. Palmer, Nicolas Chaline, Nigel E. Raine, Adam Tofilski, Stephen J. Martin Jacobus J Boomsma and Bo V Pedersen, 2005. Quantifying honey bee mating range and isolation in semi-isolated valleys by DNA microsatellite paternity analysis. Conservation Genetics 6:527–537 DOI 10.1007/s10592-005-9007-7

CHAPTER 7.
Why does a queen mate with so many males?

Caroline Otero was born poor, but before her fortieth birthday she had become one of the richest women in the world. She began her career as a dancer, but also as a courtesan and mistress of famous artists and men of power. An exhaustive list of her lovers, at least those who are known, is not the subject of this article. Let's leave aside the artists and wealthy industrialists and confine ourselves to the princes: Prince Albert I (Monaco), King Leopold II (Belgium), Tsar Nicholas II (Russia), Kaiser Wilhelm II (Germany), King Alfonso XIII (Spain), King Edward VII (United Kingdom), King Peter I (Yugoslavia), King Abbas II (Egypt) and Shah Reza Pahlavi (Iran). One explanation for Caroline Otero's exceptional list of lovers could be childhood trauma. What is certain is that collecting conquests did not cause her any problems, and that the loss of her fortune was due solely to the Monte Carlo casinos.

The queens of honey bees also mate with a very large number of males, unlike most other hymenopteran insects, which mate only once. This is the case for many species of wasps, parasitic wasps, solitary bees and bumblebees. Why do honey bees behave this way?

Originally, honey bee researchers thought that the sperm from a single drone was insufficient for a queen to lay fertilised eggs throughout her life. As it turns out, this is not the case. As we saw in the previous chapter, one drone produces well over twice as many spermatozoa as the queen can store. The queen can receive more than a hundred million spermatozoa from the drones with which she mates during a mating flight, but stores only a small proportion of them in her spermatheca. There must therefore be another explanation for this high number of matings. The researchers therefore compared bee colonies where the queen had been inseminated either with the sperm of a single drone or with that of twelve or more males. The colony whose queens had been inseminated by a large number of males were better able to discover new sources of food and to communicate them by dancing more than the others. They also produced more olfactory signals during the dance. They were also found to collect more pollen and nectar and to be more efficient in caring for their broods. The nurse bees in these colonies ate more pollen and fed more protein to the brood. We have seen that colonies with queens that have had many partners are more efficient. It seems certain that they will develop into larger colonies that will produce more swarms and drones. Other positive effects have been found for this extreme form of polyandry, which refers to mating with several males.

For example, Ding and his colleagues studied the colonisation of Australia by the Oriental honey bee. The queens of this species mate with 25 to 30 drones. The extreme polyandry of these bees presents a real advantage when colonising a new territory: an inseminated queen carries 75% of the genetic variation of the original population in the sperm of all the drones with which she has mated. As a result, only a few colonies are needed to establish a new population without a large loss of genetic variation. For the same reason, Thomas Seeley believes that queen polyandry is also beneficial to the recovery of a population if it has been affected by an epidemic.

However, the main function of extreme polyandry seems to be disease prevention. Microbiologists often cultivate pathogens in climatic chambers where the temperature is a constant 35°, because at this temperature bacterial colonies and viruses develop rapidly. The temperature of a honey bee colony's nest is also constantly maintained at 35° by the workers. This is the ideal temperature for the rapid development of eggs and larvae, but it is also the temperature at which bacteria, viruses and other pathogens develop. The Corona pandemic has shown us once again that keeping a distance between individuals reduces the risk of viral infection, but bees in a hive live in constant close contact with each other and exchange food. An infected bee can therefore easily transmit pathogens to other members of its brood. Honey bees are therefore very vulnerable to infection.

They have a series of strategies at their disposal to prevent disease and reduce cross-infection. One of these strategies is, as we saw earlier, the elimination of sick individuals through hygienic behaviour; another strategy is self-medication with antibiotics collected by the workers (see chapter 21), but, as prevention remains the best remedy, the best response of individuals against pathogens is the immune response of individual bees.

As a result of the queen's extreme polyandry, the workers have very different genetic characteristics, as they are made up of groups from different fathers. The variation in immunity within a colony increases the likelihood that some workers will be resistant to infection and that the colony will survive. Suresh Desai and Robert Currie demonstrated this by creating different types of bee colonies using artificial insemination. They exposed these colonies to parasites and pathogens. Colonies in which the queen had made a mating flight and colonies whose queens had been inseminated with a mixture of sperm from twelve drones were larger and less affected by pathogens at the end of the season than colonies whose queens had been inseminated with sperm from a single drone.

Recommended reading

Suresh D. Desai and Robert W. Currie, 2015. Genetic diversity within honey bee colonies affects pathogen load and relative virus levels in honey bees, *Apis mellifera* L. Behavioral Ecology and Sociobiology 69: 1527-1541.

G Ding, H Xu, BP Oldroyd and RS Gloag, 2017. Extreme polyandry aids the establishment of invasive populations of a social insect. Heredity 119: 381–387

Chapter 8.
The paradox of the old queen's departure.

Most insects reproduce only once and then die. Eusocial insects are the exception to this rule. The queens of termites, ants, honey bees and stingless bees live for several years and reproduce over several seasons. As a general rule, the queens of eusocial insects remain in the colony in which they began their reproductive life. In ants and termites, reproductive females lose their wings and therefore the ability to disperse. It is the fertile daughters of the queens who leave the nest and set up a new colony elsewhere. The exceptions to this rule are colonies of polygynous ants (= with several queens) and certain species of stingless bees. The other exception is the honey bee, where the queen mother leaves with a swarm and a daughter takes over the nest. Similarly, in many birds and mammals, the rule is that the mother stays at home and the daughters go elsewhere, although there are a few exceptions. In the American red squirrel, the Colombian ground squirrel and the kangaroo rat, some of the older mothers disperse and leave the territory or mound where they used to live to their daughters. In social mammals whose daughters remain in the group, the mother does not leave.

The honey bee is therefore a unique exception. Honey bees reproduce in swarms that leave the old colony. The first and often only swarm a colony produces leaves with the old queen and a daughter takes over the old nest. Why doesn't the old queen stay in the old nest and why doesn't the young queen leave with the swarm?

To answer this question, we need to determine the best possible choice for the old queen. We therefore need to calculate the expected reproductive success for that season for each of the two choices. The reproductive success of a queen is the sum of the reproductive success by daughters and by sons + the expected reproductive success of the queen for the following seasons (= the probability that she and her colony will survive the winter.

The reproductive success of a queen's daughters is the number of daughter queens that manage to become queens themselves in a colony, i.e. the number of swarms that her colony produces. In general, there is only one, and that's what we'll assume here. So there is only one daughter who gets a colony. The other young queens die. The daughter has half the queen's genes, so her relationship is 0.5. Her contribution to the queen's reproductive success is therefore 0.5 * the probability that the young queen's colony will survive the following season.

The queen has produced several thousand drones with her colony. Only a few of them will succeed in inseminating a young queen, as competition between the drones is enormous. Even drones that have successfully inseminated a queen are not sure that their genes will be used by the inseminated queen to fertilise the next generation of queens. In fact, only a small number of young queens are produced, whereas the queen has stored the sperm of 10 to 20 drones. The probability of actually using a drone's sperm is therefore less than 1. The contribution to reproductive success made by sons therefore arrives a year later (when the inseminated queen starts producing young queens), than that made by daughters and must therefore be discounted for this delay. As we don't know which drones succeed in mating and what the exact probability is that their genes will end up in a young queen, we cannot calculate reproductive success by sons correctly. We could get round this problem, because we know that honey bees invest on average equally in the production of queens and drones. This suggests that reproductive success via sons and daughters is equal. When the queen leaves with the swarm, she has already produced all the drones for the season. The same applies if she stays in the old nest. Since reproductive success via sons is the same in both decisions, we can omit it from our calculations.

As honey bee queens can reproduce for more than one season, it is important for the reproductive success of the queen that she and her colony survive the following winter. This survival rate probably depends on the decision to leave the old nest or to stay there. The old nest has the advantage of not requiring the construction of combs and the food is already there. Leaving for a new nest takes time and energy, as the bees have to search for a new nest and build new combs. This makes it more difficult for a swarm to gather enough food for the winter. It is therefore to be expected that a swarm's chances of surviving to the next season are lower than those of the old nest it has left. It is not easy to find data to verify this hypothesis. Indeed, as beekeepers do everything they can to help their bees survive the winter, we cannot use data on reared bees. Fortunately, two studies have measured these values for a population of wild bees. In the forests around Ithaca, New York, the probability that an established colony would still be there after winter was 0.80 in the 1970s and 0.84 in the 2010s. The probability of a swarm surviving its first winter was 0.29 in the 1970s and 0.23 in the 2010s.

The queen must choose so that there are as many copies of her genes as possible in the bee population during the following season. These copies are formed by the sum of her own genes + half of her daughter's genes (the other half of the daughter's genes come from the drones that managed to mate with the young queen).

The sum to be maximised is therefore 1.0 times the probability that the queen herself will survive plus 0.5 times the probability that the daughter will survive. If we use the probabilities found by Seeley in his study of wild bees, we obtain for the 2017 data: If the queen stays 1 * 0.84 + 0.5 * 0.23 = 0.955. If the queen leaves 1* 0.23 + 0.5*0.84 = 0.65 It is therefore in the queen's interest to stay at home and let her daughter leave with the swarm, which is also the case for the previous experiment from the 1970s.

The forests near Ithaca, New York, are not a very rich habitat for bees, and the winters there are long and cold. Would the result be different in a milder habitat? In a habitat where the probability of a swarm surviving the winter is as high as that of the established population, the values for leaving and staying are similar. Even in this case, leaving would not allow the queen to increase her reproductive success more than staying.

The conclusion is that the survival rate of the swarm cannot be the main reason why the old queen leaves the nest with the first swarm. Only if the old queen had a lower survival rate if she stayed an extra year could the prediction be reversed. Most of the colonies established in Seeley's studies produced a swarm (95% in the 1970 experiment, 87% in the 2010 experiment). These colonies therefore had a new queen. The question now is whether the survival of the established colonies would have been as high if the old queen had stayed behind.

The departure of the old queen and the takeover of the nest by a young queen have enormous consequences for the genetic composition of the colony. Although the young queen has half the genes of the old queen, she mates with 10 to 20 drones that are not related to the old queen, ensuring that the workers in the colony all have fathers that are different from those of the old queen. If the old queen had stayed, the genetic composition of the colony would have remained the same. We could explain the departure of the old queen and the takeover by her daughter if the radically altered genetic composition of the colony was important for the survival of the colony in the following season. This could be the case if, in a nest where the old queen remains, the probability of disease is higher.

The high temperature of the nest, the numerous contacts between bees in a hive and the frequent exchanges of food mean that pathogens can spread rapidly in a hive. The honey bees' first line of defence against pathogens is the antibiotic substances contained in the plant resins collected by the bees to make propolis. The nest of wild honey bees is surrounded by an envelope of propolis. As nectar and pollen also contain antibiotic substances, this food is also a pharmacy. The honey bee's main weapon and last line of defence is its own immune system.

Bacteria, fungi, microsporidia and viruses have much shorter generation times and much larger populations than bees. As a result, they often produce new mutants, some of which could potentially break down existing immune defences. Pathogens present in a colony can adapt to the colony in which they reside for an entire season. During this season, they have had a large number of generations. In each of these generations, the pathogens best able to evade the bees' immune responses were at an advantage.

As long as the old queen is present, the genetic make-up of the bee colony does not change. But pathogens can continue to evolve and become more virulent during this period. Selection within a colony over a relatively short period favours the most virulent pathogens because they produce more offspring than less virulent variants. When the old queen leaves and is replaced by a young inseminated queen, new alleles enter the colony, potentially cancelling out the adaptation of the pathogens to the colony and helping the bees to resist the pathogens present. The old queen migrates to a new environment, escaping many pathogens.

If the old queen remained, the pathogens could continue to adapt and become even more virulent over an entire season, threatening the survival of the colony. This hypothesis provides a possible explanation for the old queen's departure. The prediction is that if the old queen stays, the colony's chances of survival the following winter will be lower than if a young queen takes over. This prediction can be tested experimentally. There are currently no clear data known in the literature to support the hypothesis, although in a large experiment by Ralph Büechler et al (2014) 85% of colonies died in the second season and survival curves steepen to low after the first year. In Seeley's experiment, the chance of an established population producing a swarm was 0.95 in the 1970s and 0.87 in the 2010s. Seeley did not distinguish between established colonies that produced a swarm and those that did not. But even if we knew the survival rates of colonies that didn't swarm, we couldn't properly compare the two categories, because of the small number of colonies that didn't swarm and because colonies that don't swarm differ from those that do. Colonies that do not swarm are too small to swarm or have insufficient stocks.

However, the hypothesis that the departure of the old queen is a means of giving the remaining colony a greater chance of surviving infectious diseases is the most likely explanation for this exceptional behaviour. Thus, the departure of the old queen could be part of a series of unique adaptations that honey bees have to protect themselves from pathogens. These other unique adaptations are: the mating behaviour that results in a panmictic population structure, the extreme polyandry of the queen and the extremely high frequency of recombination. Young queens mate with 10 to 20 different drones and use the sperm of all these drones to fertilise their eggs (see Chapter 9). This ensures that the workers in a population are the offspring of many different fathers and therefore have different genetic traits. This can limit the spread of pathogens in a colony. The panmictic structure of the population and the extremely high frequency of recombination combine to form a double-edged sword in honey bees' fight against new bacterial or viral infections. Thanks to mating behaviour, new rare alleles can be recruited from a large population, which can then be combined into new genotypes by recombination with useful alleles from other genes. In this way, honey bees can compensate for slower reproduction and lower population densities than bacteria and viruses.

The fact that there is selection within a colony for increased virulence of pathogens does not contradict the theory that vertical transmission favours the evolution of a-virulence, while horizontal transmission (the spread of disease between unrelated individuals) favours the evolution of virulence. Selection in favour of reduced virulence occurs in the longer term through selection between colonies: colonies containing highly virulent pathogens will die out, while those containing less virulent pathogens will survive.

Preventing swarming, either by breaking the cells of young queens or by cutting off a queen's wing, could therefore result in a higher risk of infectious diseases for the colony.

Recommended reading

Thomas Seeley, 1978. Life History Strategy of the Honey Bee, *Apis mellifera*. Oecologia (Berl.) 32, 109-118.

Ralph Büchler, Cecilia Costa, Fani Hatjina, Sreten Andonov, Marina D Meixner, Yves Le Conte, Aleksandar Uzunov, Stefan Berg, Malgorzata Bienkowska, Maria Bouga, Maja Drazic, Winfried Dyrba, Per Kryger, Beata Panasiuk, Hermann Pechhacker, Plamen Petrov, Nikola Kezić, Seppo Korpela and Jerzy Wilde. 2014. The influence of genetic origin and its interaction with environmental effects on the survival of *Apis mellifera* L. colonies in Europe. Journal of Apicultural Research 53(2): 205-214

Jacques JM van Alphen and Robin Owen, 2024. The paradox of the old queen leaving. Chapter in: Hymenoptera - Unanswered Questions and Future Directions. London, IntechOpen publishers. In Press.

CHAPTER 9.
The arms race

Further proof of the need to vary immune responses in honey bees is provided by another extreme adaptation. This is a process that is part of egg formation. During this process, the corresponding chromosomes pair up. Sometimes breaks in the chromosomes are then repaired, but in such a way that the broken parts swap places and each piece of a chromosome is attached to the other. The chromosomes therefore change composition. After this exchange, called "recombination", each of the two chromosomes is found in an unfertilised egg. Recombination thus creates new genetic variants and can result in favourable combinations on the same chromosome. Recombination can also separate favourable combinations of genetic variants. As a result, there is an optimal amount of recombination. The costs and benefits of recombination are subject to natural selection. The probabilities of chromosome breakage, the locations of breakage and recombination are themselves heritable. They differ between species and even between populations of the same species. The honey bee has the highest frequency of recombination in the entire animal kingdom. It has more recombination than bumblebees, and the European honeybee more than the South African subspecies. This is why the extremely high frequency of recombination in honey bees should more than offset the risks and be beneficial. Evolutionary biologists believe that recombination is very important in the fight against bacteria and viruses, as it increases the chances of the emergence of a resistant variant. Compared with the organisms they attack, bacteria and viruses have huge populations and a very short generation time. This means that they often generate new mutations and can evolve rapidly. Recombination is the ideal solution for organisms with smaller populations and longer generation times. With recombination, they can create a large number of new genetic variants with each generation, enabling them to combat new virulent mutants of their pathogens. But because recombination has also negative effects (if favourable combinations are disrupted), natural selection limits the frequency of recombination to what is strictly necessary. Only strong selection by pathogens can lead to an extremely high frequency of recombination. The extreme polyandry and very high recombination frequency of honey bees show that it is even more important for them than for other animals to constantly create new genetic variants. As drones and queens travel great distances to meet at the former's congregation sites (see chapter 6), genetic traits are mixed again and again over large areas within large populations. Extreme polyandry, the very high frequency of recombination and the mixing of genetic traits of bees from a vast area are three essential characters which, together, help bees to cope with the infections to which they are vulnerable because they live in the promiscuity of heated nests. Bee breeding, using isolated mating stations or artificial insemination, takes little account of this insurance against new variants of pathogens. The ravages caused in Europe by the arrival of the varroa mite and associated viruses are a consequence of this. Selection characteristics of bees such as high honey production, low propensity to swarm, lower aggressiveness and specific hygienic behaviour in the face of varroa mites involve complex behavioural traits. Many genes are involved in the inheritance of these characteristics. Most of these genes only appear when an individual has inherited two identical alleles, one from the mother and one from the father. Such genes are known as "recessive". Bees selected for the characteristics mentioned above therefore have two identical copies of the recessive genes in question - they are "homozygous". Genes close to those selected on the chromosome also take part in the selection process and therefore also become homozygous. Recombination can only be effective if the alleles exchanged between the chromosomes are different. Homozygosity reduces the production of new genetic combinations by recombination and therefore the chances of developing resistance to new pathogens. Selection based on insemination with the sperm of a single drone eliminates the weapons available to bees against new pathogens, namely the mating structure of the bee population and the queen's choice of mate (see chapter 6 and 7) and polyandry, while reducing the effect of recombination. The selection of bees with high honey production, low swarming tendency, low aggression and varroa-specific hygienic behaviour is entirely possible, but as well as leading to the homozygosity mentioned above, it also leads to the loss of rare genetic variants. In a large population of freely mating bees, most genes have several variants. In the honey bee, the number of alleles of a given gene (alleles) can range from 1 to 37 or more. Some alleles are common, but many are quite rare. It is likely that the common alleles have become so through natural selection. These alleles are therefore currently useful to bees. This does not mean, however, that the rare alleles are useless. Who knows whether, in a changing world and under new conditions in the future, these rare alleles could become important for adapting the bee population to the new circumstances? If we choose a few colonies from such a large population to begin selection, we obtain a small sample of this population.

It is then likely that common genetic variants will be chosen in this sample. This phenomenon is repeated ad infinitum when the selection is repeated in subsequent generations. Even if we tried to keep genetic variation as high as possible, by creating a number of parallel breeding lines which could then be crossed, we would still lose rare alleles through selection in each of the lines. Finally, once the selection lines have obtained the desired characteristics, the beekeeper wants to keep them: so, they have to mate in a small population of selected bees in a mating station. This ensures that rare alleles do not reappear. The bee breeder suffers no loss. During selection, he has ensured that the bees selected are always highly heterozygous by working with a number of different selection lines. The alleles in his selection lines have become common in the original population through natural selection, so they are not bad variants. His bees therefore appear to be very healthy. However, when the population is threatened by a new disease or parasite, the absence of rare alleles becomes problematic. The genetic variation needed to develop resistance to these new diseases is then lacking in the selected population. Indeed, Themudo et al. 2020, showed that genetic diversity of European honey bees has been declining during the 20th century. This could explain why 'pedigree' bees have not developed resistance to the varroa mite and its associated viruses for 40 years, whereas populations of 'natural' and free-living bees did after a short period of time.

Recommended reading

Gonçalo Espregueira Themudo, Alba Rey-Iglesia, Lucía Robles Tascón, Annette Bruun Jensen, Rute R. da Fonseca & Paula F. Campos, 2020. Declining genetic diversity of European honey bees along the twentieth century. Scientific Reports (2020) 10:10520.

Jack Hassett, Keith A Browne, Grace P McCormack, Elizabeth Moore, Native Irish Honey Bee Society, Gabrielle Soland & Michael Geary. A significant pure population of the dark European honey bee (*Apis mellifera mellifera*) remains in Ireland. Journal of Apicultural Research 57: 337–350 doi.org/10.1080/00218839.2018.1433949.

Akira Sasaki and Yoh Iwasa, 1987. Optimal Recombination Rate in Fluctuating Environments. Genetics 115: 377-388.

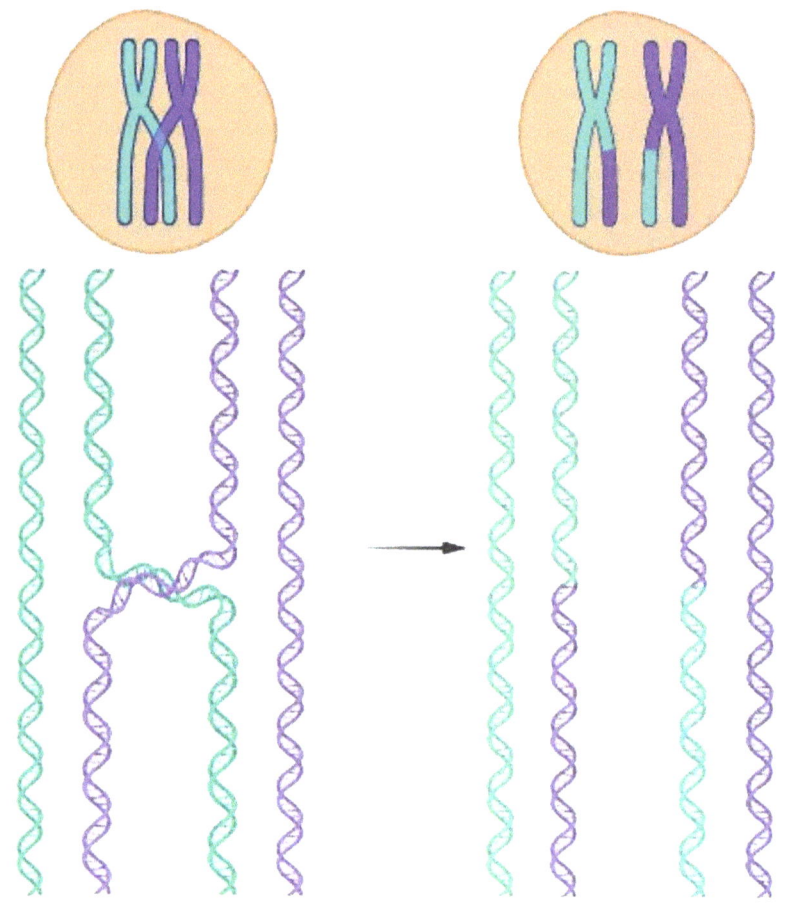

Recombination

Chapter 10:
Why so many drones and so few queens?

Charles Darwin was a great original thinker, but he wasn't always convinced of it himself. He was never sure that he had understood anything completely. In 1871, he wondered why many animals gave birth to approximately the same number of females as males. It is remarkable that he asked himself this question. His contemporaries considered the sex ratio to be a fixed fact. Darwin asked this question because he thought that sexual selection would take place more quickly if there were more males than females, and that there would therefore be more competition between the males. Sexual selection is a form of selection that results from competition between males for females and the choice of mates. This was an important subject for Darwin, as he could not explain the evolution of male ornaments, such as deer antlers and peacock tails, by natural selection. So, he looked at the sex ratios of as many animals as possible, but to his surprise, he found no great disparity between the numbers of the two sexes. He then tried to understand why, and how natural selection affects the sex ratio.

Fifteen years after The Origin of Species, he published The Descent of Man and Selection in Relation to sex. In this book, he explains how he thinks natural selection can lead to the birth of an approximately equal number of men and women.

His reasoning was correct. Darwin then began to doubt the accuracy of this theory and, three years later, in the second edition, he revoked it, saying: "*I now see that the problem is so complicated that it is safer to leave its solution to the future*". We now know that Darwin's theory was correct. He never properly explained why he retraced his steps.

It was not until 1930 that the theoretical biologist and mathematician Ronald Fisher rediscovered Darwin's theory. He reasoned, like Darwin before him, that natural selection leads to an equal ratio of sons and daughters when sons can provide as many grandchildren as daughters. After all, suppose male births are less frequent than female births. In this case, a son has a better chance of mating than a daughter, and can therefore hope to have more offspring. Parents who have a genetic predisposition to produce mainly sons will therefore have an above-average number of grandchildren. As a result, male disposition genes will increase and males will be born more often. When the sex ratio approaches 1:1, the advantage of producing sons disappears. The same reasoning applies when female births are less frequent than male births. Natural selection therefore leads to an equilibrium of an equal number of sons and daughters. However, Fisher's reasoning is only valid if all males and females have the same chance of mating. In other words, when all the members of the population can meet. In chapter 6, we saw that honey bees meet to mate at drone congregation areas, and that animals from hundreds of colonies over an area of up to 150 square kilometres meet there. In the case of honey bees, the condition that a population should be well mixed is thus fulfilled.

Bee colonies produce around 20,000 drones per season and only a few young queens. So they don't seem to fit in with Darwin's and Fischer's theory. But Fisher mentions another important condition, which was not mentioned by Darwin: raising a son must cost as much as raising a daughter. This is generally true, but not always. Bill Hamilton applied Fischer's reasoning to honey bees. Not only did he think that the cost of rearing a young queen had to be taken into account, but also that, as bee colonies reproduce by dividing into swarms, the cost of producing the swarm had to be added. Such a swarm is made up of around 12,000 workers. The prediction of Fisher's theory is therefore that the cost of rearing the queen and such a swarm of workers must be equal to the cost of producing drones. Robert Page and Robert Metcalf studied this question by examining the production of drones, queens and swarms in a population. They found that the investment was roughly the same for both sexes.

Finally, we return to Darwin's idea that if many more males than females are born, we can expect strong sexual selection. The honey bee is a good example of this. At drone gathering sites, thousands of drones compete to mate with just a few queens. Only the strongest and fastest drones succeed. Drones that have been attacked by varroa mites or viruses during their development are weaker and less able to fly. drones are also more susceptible to infection by viruses and bacteria because they have only one set of chromosomes. Strong sexual selection on drones is therefore of great importance for the resistance of honey bees to parasites and pathogens. This was also demonstrated in a breeding programme in which bees were selected for resistance to varroa mites: the resistant queens that were eventually bred were used to create colonies that produced only resistant drones. The use of these colonies proved much more effective than the use of resistant queens alone in spreading resistance. Apparently, resistant drones are very successful in an environment where there are few or no resistant bees.

Darwin thought of sexual selection mainly in terms of the colourful ornaments of birds of paradise, the tails of peacocks and the antlers of deer.

He had no idea that sexual selection sometimes involves characteristics that also offer an advantage in natural selection, such as being able to harvest good food or being resistant to varroa mites. He expected an unequal sex ratio in the animals whose males were adorned, but this did not turn out to be the case. It's a pity he didn't think of the honey bee, because then he would have had a good example of an animal with an unequal sex ratio and with sexual selection on drones, which helps natural selection because only healthy drones succeed in mating. An example that fits both his theory of the sex ratio and that of sexual selection.

Recommended reading

Charles Darwin, 1871. The descent of man, and selection in relation to sex. London: Murray. [1st ed.]

R. A. Fisher, The Genetical Theory of Natural Selection (Dover, New York, 1930).

Hamilton, W. D. (1967). Extraordinary sex ratios. Science 156, 477–88.

Chapter 11:
The language of bees: waggling and trembling

Aristotle wrote extensively about bees in his History of Animals. A closer look at the text of Book IX, Volume 40, reveals the following sentences: (1) *From dawn they are silent until one bee buzzes two or three times, waking the others, and then they all fly in groups to their work.* (2) *Each bee that returns to a flower is followed by three or four companions.* These are the earliest known quotations on communication between bees, and they show the extent to which Aristotle was able to perceive and interpret what he saw. Between the observation that bees communicate and the answer to the question of how they do it, there is a period of almost 2,300 years. We owe the decoding of bee communication to Karl von Frisch.

Karl von Frisch (1886-1982) was the son of a doctor. He had a great love of animals from an early age. At his father's insistence, he studied medicine, but gave it up after two and a half years and decided to study biology. For his thesis, he studied colour changes in fish. Going from the colours of fish to those of flowers was only a small step. At the time, many biologists thought that bees and other insects were colour-blind. Von Frisch could not believe it, because, he thought, the bright colours of flowers could only be understood as an adaptation to insects that find the flowers they pollinate by their colours. So in 1912, he began experiments to show that bees could indeed see colours.

On a table outside, he placed a coloured paper between papers of different shades of grey, and on top he placed a small glass dish filled with sugar syrup. Bees from a nearby hive were trained to recognise the colour, demonstrating their ability to distinguish colour from different shades of grey. Between tests, the bowls were empty. He noticed that scout bees came only sporadically to the empty dishes, and those that did come naturally flew back to the hive without success; the feeding table remained deserted afterwards. On the other hand, if an explorer found the full dish and returned home successfully, a whole group of bees would be back at the feeding site within minutes.

Had the scout brought his findings back to the hive? This question was the starting point for a new line of research. To observe the behaviour of returning scouts, Von Frisch placed a small colony in an observation box with glass windows. Next to it, he placed a dish with sugar syrup. He marked a large number of bees with coloured dots, so that he could recognise them individually. What he saw in the observation box was astonishing: even before the returning bees had emptied themselves of their load of sugar syrup, they were running around the comb in small circles, alternately to the right and to the left. This circular dance ensures that marked bees folowing a returning bee are making an excursion towards the feeding site. In the next experiment, Von Frisch offered food to the bees at two different locations. The different feeding sites were visited by two different groups of bees, each following a different scout bee. Von Frisch then repeated the experiment, but instead of two dishes containing sugar syrup, he placed a bunch of lime blossom in one place and a bunch of black locust blossom in the other. After a pause during which the flowers were removed, only the black locust flowers were put back and not the lime blossoms. In this case, the scouts who had found the returned black locust flowers only recruited bees that had already visited the black locust flowers; previous visitors to the lime flowers showed no interest. Apparently, the scouts brought with them the scent of the flowers they had visited and the workers looked for a place where this scent was present. When Von Frisch then experimented with dishes of sugar syrup to which mint oil had been added, he was able to confirm that the scouts communicated the smell of the food source to the bees in the hive. The richer the food source, the more enthusiastically the scouts danced and the more bees were recruited. Von Frisch published these experiments in 1923 and believed he had understood the language of bees. The scouts recruited other workers with their dancing rounds and received information about the scents to look for outside.

It was not until twenty years later that he decided to repeat this type of experiment and quickly realised that he had missed the most important thing. During the experiments described above, Von Frisch had already noticed that some of the bees in his hive were performing a different type of dance, waggling their abdomens, but these bees had never come to his saucers of sugar syrup. Von Frisch originally thought that the round dance communicated the location of a nectar source and the waggle dance the location of a pollen source.

In the new series of experiments, saucers containing sugar syrup were proposed at a greater distance from the nest. To his great surprise, the scouts who had found such dishes and returned to the hive did not do a circular dance but a waggle dance. The idea that the dance was recruiting bees to a pollen source was immediately dismissed; it was all a question of distance. His error was due to the fact that, in previous experiments, the bees with pollen always came from further away than those visiting the dishes with syrup.

There were then experiments in which the distance between the saucer containing the sugar syrup and the observation box was varied. It was found that longer distances were expressed by longer waggling times. The sweeter the sugar syrup, the more lively the bees danced and the more often the dance was repeated. These experiments also showed that the round dance was in fact a distinct form of the waggle dance, with an extremely short waggle time because the feeding site is close to the hive.

The waggle dance seemed not only to indicate distance, but also the direction of the target. In fact, in the observation hive, the bees coming from the same feeding site made their waggling paths in the same direction, whereas these paths were directed differently for bees coming from other directions. It was also found that the direction of waggling bees coming from the same feeding site changed over the course of the day. As the hours passed, the direction changed at the same angle as that taken by the sun. In this way, the waggle dancer shows the other bees the direction of the target in relation to the position of the sun. In this way, the dancer scouts are able to provide the other bees with information about the direction and distance they need to travel to reach a feeding site, the quality of that site and the type of flowers involved. What is fascinating is that the angle between the position of the sun and the dancer's trajectory towards the target is expressed in the darkness of the hive, on the vertical surface of the comb. The bee thus translates the angle between the direction of the sun and the direction of the food source found into the angle that the direction of the waggling dance makes with the direction of gravity. The dance of the bees is therefore a symbolic representation. A simple and elegant experiment demonstrated this: Von Frisch tilted his observation box so that the comb on which the bees were dancing was horizontal. If the bees could see the sky, then the waggle dance took place exactly in the direction of the food source. If the sky was then shielded, the bees danced confusedly in all directions and the bees following the dances then had no idea where to fly. As soon as the hive was returned to its normal position, with the comb vertical, the bees danced in the direction of the food source.

So, the bees orientate themselves in the dark by gravity. But if they can see the sky, they orientate themselves towards sunlight. The bees were then suspected of perceiving the direction of polarisation of the light and thus always knowing where the sun was. The following experiment confirmed this. When a rotating polarising filter was placed over the comb so that the direction of the polarised sky light did not change, the bees continued to dance correctly. However, when the polarising filter was turned to the right or left, the bees' dancing direction changed to the right or left with corresponding angular values.

When the bees return home after finding a good source of food at the end of the day, the sun is in a very different place; yet they return without hesitation to the food source the next morning. The bees seem to have an internal clock and know how the position of the sun changes throughout the day. Von Frisch was also able to demonstrate this through experiments.

For all this research, Von Frisch was awarded the Nobel Prize in 1973, but not everyone was happy about it. The fact that bees could transmit detailed information to their hive mates was difficult to accept for some scientists and many non-scientists at the time. Language was considered the exclusive domain of humans. Yet there are concrete experimental data showing that even an insect, can inform its hive mates of the location of a resource.

Decoding the waggle dance has not yet revealed the whole story. The function of the waggle dance is to recruit other bees to exploit a new food source. Bees returning from a feeding flight with a load of pollen or a crop full of nectar transfer the food to the young bees in the nest who carry out domestic tasks. On their return home, they must find a bee to take charge of the load and store it. The bee that fetches the food can then return quickly to the food source. If many bees are recruited to collect the food, many bees are also needed to take charge of and store the food brought in. To match the food supply with the ability to transfer and store food, the bees also need to communicate. Suppose that temporarily too many bees have been recruited to fetch food: the supply of food by recruiting extra bees has become so great that when they return, they have to wait and look for a bee to take over. A signal is then needed to ensure that more bees are available to take over and store the food they are looking for. Or, if this extra capacity is not available, then a signal is needed to encourage fewer bees to go and look for food. Similarly, if a food source suddenly dries up, for example because a farmer mows a meadow full of flowers, or because of sudden bad weather, a signal is needed to encourage fewer bees to go in search of food. In 1923, Von Frisch had already noticed that bees returning to the hive sometimes performed a different type of dance. In this dance, not only does the bee's abdomen vibrate, but the whole bee shakes and sways on the comb with its front legs raised. Von Frisch compared this behaviour to a St. Vitus dance. The dance was not restricted to a certain direction, and was sometimes aimed at the bees doing the waggle dance, but sometimes also at the bees in the nest that came to take over the harvested food from from foraging bees. Von Frisch therefore saw no clear function for it and concluded that there probably wasn't one.

In 1948, Martin Lindauer published experiments in which he had fed bees with a sugar solution to which he had added a lot of salt. The bees that had to take up the food did so only sparingly, so the bees that went to collect the food had to wait a long time before being relieved of their salt load and being able to fly away again. Lindauer found that bees waiting to collect their food often did the tremble dance and interpreted this as a reaction to the poor quality of the food. In 1987, Thomas Seeley conducted an experiment in which the bees that came to take over the food were removed at the end of the session, in the hope that there would then be far fewer bees taking over the food the next morning. This proved to be the case, and the bees that returned with a load of food therefore had to wait longer before being relieved of their load. Although the experiment had a different purpose, Seeley often saw the waiting bees doing the tremble dance. This gave rise to the idea that the tremble dance is a signal that prompts more bees to help take over and store the food brought in, and fewer bees to leave in search of food. Seeley conducted further experiments to prove this. This prompted Martin Lindauer to look again at his earlier study with the salted sugar solution. With his student Wolfgang Kirchner, he repeated his old experiments, but this time he also recorded the waiting times of the bees that brought the food. In this way, they were able to confirm that the trembling dance is a reaction of the foraging bees to the long wait before taking up their cargo.

The tremble dance is therefore a feedback mechanism that prevents that more food is collected than can be handled. The message of the tremble dance seems to be: there's more nectar coming in than we can handle. For the nest bees, the meaning is: I have to switch tasks to taking nectar from the bees that are bringing it. For the waggle dancing bees, the message is: I must stop recruiting other bees. The combination of the waggle dance and the tremble dance enables a population of bees to react quickly to changes in the food supply. It has taken three generations of researchers to decode the rich language of bees.

Recommended reading

Karl von Frisch, 1973. Decoding the language of the bee. Nobel Lecture. December 12, 1973.

Martin Lindauer, 1948. Über die Einwirking von Duft und Geschmackstoffen sowie anderer Faktoren auf die Tänze der Bienen. Zeitschrift für vergleichende Physiologie 31: 348- 412.

Thomas D. Seeley, 1995. The Wisdom of the hive. Harvard University Press, Cambridge Massachusetts.

Chapter 12:
Choosing a future home

We would have been very close to never hearing about Martin Lindauer. This clever farmer's son was conscripted into the German army against his will in 1939 and found himself in a company sent to the Eastern Front. There, on 11 July 1942, he was hit by Russian shrapnel that destroyed a nerve in his arm. His arm was largely unusable. He was discharged from further service and returned to Germany, while the other 156 members of his company were sent to Stalingrad, where all but three were killed.

Lindauer decided to study biology and, in the spring of 1945, began his doctoral research on honey bees with Karl von Frisch as his thesis supervisor. The previous summer, Von Frisch had made his revolutionary discovery, for which he was to receive the Nobel Prize: honey bees use dance behaviour to inform their hive mates of the direction and distance of a rich food source. Von Frisch wanted to investigate this further, so he asked Lindauer to study how this communication is influenced by the smell and taste of food. Lindauer carried out his doctoral research independently in Munich, while Von Frisch worked temporarily in Graz, Austria, as the Munich Zoological Institute had been destroyed by bombing. Shortly after the war, it was virtually impossible for Germans to travel abroad, so contact between supervisor and student was rare. When Lindauer finished in 1946, he sent his thesis to Austria. Von Frisch was very impressed by the quality of the research and gave Lindauer a position as an assistant.

Lindauer had a keen eye for the unexpected. In the spring of 1949, he accidentally found a swarm from one of the hives at the Zoological Institute. Before the beekeeper arrived to move the swarm to a hive, he noticed that bees were dancing on the surface of the swarm. He also noticed that these bees were not carrying loads of pollen or pausing to unload nectar. Instead, they danced without pausing. He noticed that some of the dancing bees were covered in red brick dust, chalk or soot. He therefore thought that they were scouts looking for a new home for the swarm. This observation gave rise to a new research project. Thanks to Von Frisch's earlier research, Lindauer was able to translate the information from the waggle dances into directions and distances. Could he use this information to track the swarm's decision-making process? At first, the scouts used their dances to announce all sorts of locations in different directions and at different distances. Sometimes, up to 25 different nesting sites were announced. The dancing bees thus recruited new scouts, who then went to take a look at the advertised nesting site. If they were enthusiastic, they returned to the swarm to dance and recruit scouts themselves. Poorer locations produced fewer new scouts and when they returned, they danced less enthusiastically and less often. Little by little, more and more scouts danced for fewer and fewer places, until finally almost all the scouts were dancing for the same place. It was as if the bees were deciding democratically where they would live. Lindauer suspected that the bees convinced each other that one place was better than another. He thought that the bees who had found worse places were then recruited to look at better places and that by comparing the places they changed their minds. But he wasn't entirely sure, as he had observed on several occasions that bees who had previously recommended a less good place had stopped dancing for it without having visited a better place in the meantime. He also thought that the bees only left for their new home when all the scouts were unanimous in their decision, but he wasn't sure about that either. He published his observations and suspicions.

More than twenty years later, a young Harvard doctoral student, Thomas Seeley, read Lindauer's publications and found in them the inspiration for his doctoral research on the choice of habitation in bees (chapter 12). Twenty years later, with his students and his friend Kirk Visscher, he decided to continue his research into decision-making in swarming bees. An important question for him was whether the decision-making process observed by Lindauer actually leads to the best choice. After all, when the best house is only discovered late in the decision-making process, many bees have already been recruited for less good houses.

Lindauer thought, on the basis of his observations, that a scout would dance more intensely for a better nesting cavity, but he had never done any experiments to prove it. In order to make the right choice between different nesting cavities, it is important that there is a relationship between the quality of the potential dwelling and the intensity with which it is advertised. Seeley allowed swarms to choose between a good and a less good nesting cavity and found that scouts who found the good nesting cavity danced more often and for longer than those who found the less good cavities. Apparently, scouts have an innate sense of what is good and what is not and don't need to learn this through experience.

Thomas Seeley also wanted to know whether Lindauer's idea that scouts change their preferences by comparing places was correct. He investigated this phenomenon by following individually marked scouts in a decision-making experiment, in which they could choose between a good and a poor nesting cavity. The observations revealed an unexpected pattern. Most of the scouts visited only one nest cavity. If they returned to the swarm after discovering it, they danced for a relatively long time and intensely. If they visited the nest cavity again afterwards and danced again on their return, the dance was shorter and less intense. After each return to the swarm, the dances became shorter, until the scout stopped dancing on his return and rested on the swarm. For the scouts in the good and medium nesting cavities, the motivation to dance faded at the same rate. As the scouts from the poor cavity were less motivated to dance on their first return, they stopped dancing earlier than the scouts from the good cavity. So, the bees did not do what Lindauer expected. They do not compare cavities and they do not have to convince each other. The scouts motivate other scouts to visit the cavity they have discovered. These new scouts in turn motivate other scouts to do the same, and so the number of scouts dancing to advertise a particular cavity increases over time. Scouts who dance to advertise a mediocre cavity dance for less time and recruit fewer new scouts, who in turn recruit fewer new scouts. As a result, the fraction of dancers dancing for the best cavity increases and the fraction dancing for the mediocre cavity decreases. Even if a good nesting cavity is discovered late in the decision-making process, this behavioural mechanism can still lead to its selection. The mechanism can therefore lead to the best possible choice.

    The scouts work together to find as many potential nesting cavities as possible. Then, through the democratic process described above, they decide which is the best option. If most of the scouts dance for the same option, the decision can be made. It is unlikely that all the scouts dancing will know from each other which nest cavity they are dancing for. The question is therefore how they arrive at an unambiguous decision. This is not done by counting the dancing scouts. That would be like determining the outcome of an election by comparing the scale of the various parties' election campaigns. So there has to be a vote first. To find out, Thomas Seeley and Kirk Visscher carried out some ingenious experiments. They suspected that the decision would be taken in the future home. Firstly, they let the swarms choose between two equal nesting cavities of good quality. As they had hoped, no consensus emerged. Half the scouts danced for one cavity and half for the other. After a while, the swarm left anyway. Half the swarm went in the direction of one hive and the other in the direction of the other. In the air, the bees sensed that something was wrong. The queen landed and the bees formed a swarm around her again. The experiment showed that the scouts in a swarm do not wait for unanimity before making a decision. In a subsequent experiment, the swarms were tested twice. Once with five identical nesting holes of good quality that were installed right next to each other. Another time with just one of the five nesting holes. As in the first case the scouts were spread over five different nesting holes, there were never many bees present in and around a hive at the same time. On the other hand, there was a rapid increase in the number of scouts in and around a single hive. When the scouts were spread over five different hives, it took more than twice as long for the swarm to start moving. The results of the experiment are consistent with the idea that the decision is made when the number of scouts in the new home has exceeded a certain number. Seeley believes this to be around 75 scouts, of which around 30 need to be present at any one time.

So, a small group of scouts makes a decision for a swarm, which consists of a queen and 10-12,000 workers. This poses an interesting problem. Only the scouts know where the new home is. How do they get all the other bees there?

The first problem is that all the bees in a swarm have to move at the same time, and so take off at roughly the same time. The temperature at the heart of a swarm of bees is 35^0C, but only 17^0C in the outer layers. All the bees need to be at 35^0C to fly. When the bees of the future hive have reached quorum, they return to the swarm and run excitedly over the other bees. From time to time, they grab another bee and press their thorax against it, emitting a shrill squeak. As more and more bees adopt this behaviour, more and more squeaks can be heard. Together with Jürgen Tautz, Thomas Seeley was able to demonstrate that this behaviour encourages the bees to warm up to 35^0C by vibrating their flight muscles; this is how the swarm prepares to leave. When the swarm is warm, the scouts give the signal to leave. Lindauer had already seen that they do this by running over the swarm with their wings spread.

Then the swarm takes off and the scouts have to make sure that the swarm is heading in the right direction. Lindauer had observed that in a swarm on the move, there are a few hundred bees that fly faster than the others and therefore go towards the head of the swarm. They then fall back onto the tail of the swarm and set off again at high speed. The other bees follow the direction of flight of the fast-flying bees.

In 1955, it was still not technically possible to check whether the fastest bees were actually pointing in the direction of the new cavity. This only became possible in 2006, thanks to high-speed cameras and advanced image analysis. Thomas Seeley and Madeleine Beekman were able to rule out another possible explanation, namely that the scouts were leaving an olfactory trail that would then be followed by the other bees in the swarm: if they blocked the scouts' olfactory glands, the swarm would simply arrive at its destination. Madeleine Beekman then had moving swarms cross a busy thoroughfare used by bees from other colonies collecting nectar as part of an experiment. These nectar collectors also fly at high speed, so the moving bees didn't know which one to follow. Of the six swarms she examined, five did not reach the new nest cavity. This was definitive proof that swarms follow fast-flying bees. The enigma that began with a simple observation in 1949 was thus solved after almost 60 years of research.

Recommended reading

Martin Lindauer, 1955. Schwarmbienen auf Wohnungsuche. Zeitschrift für vergleichende Physiologie 37: 263-324.

Thomas D Seeley 2010. Honeybee Democracy. Harvard University Press, Cambridge Massachusetts.

T Latty, M. Duncan and M. Beekman, 2009. High Bee traffic disrupts transfer of directional information in flying honeybee swarms. Animal Behaviour 78: 117-121.

Chapter 13:
The ideal home: not too small, not too big, well insulated and well located.

In the early summer of 1963, eleven-year-old Thomas Seeley saw a swarm of bees flying near a large black walnut tree near his parents' house. Intrigued by the buzzing sound, he followed the swarm as it made its way through a hole in the trunk. After a few minutes, the swarm had disappeared. These were wild bees whose nest Thomas had found, and he returned there often that summer to see what was going on. Little did he know then that he would devote a large part of his life to the study of wild bees. He lived in a small valley called Ellis Hollow, a few miles east of Ithaca in upstate New York. It's a wooded area with a lot of unspoilt nature, a good environment for a boy who is interested in all living things. When he was in the last class of high school, Thomas built a beehive with which he captured a swarm of bees. That's how he got his start as a beekeeper. After high school, he decided to study chemistry as a basis for his subsequent medical training. During the summer holidays, he worked as an assistant at the Dyce Laboratory for Bee Studies at Cornell University, in his home town. It was there that he realised that biology was more in his heart and he chose to go to Harvard to do doctoral research on bees under the supervision of Bert Hölldobler.

    He chose the Dyce Lab as the basis for his fieldwork. While working there, he had met Roger Morse, the professor of beekeeping, who now supervised him informally. Seeley had read Martin Lindauer's publications (see Chapter 12). He now wanted to know what features of a nesting cavity bees preferred. So he began by describing in detail the nests inhabited by wild bees in the woods near Ithaca. He offered a reward for information about wild bee nests. Ten days later, he had information on 36 nests in hollow trees around Ithaca. On 33 of these nests, the height and direction of the opening were noted, among other things. Subsequently, 21 were cut down and the piece of trunk containing the nest was brought to the university to be measured. Remarkably, the average volume of the nest cavities was only 47 litres, well below the volume of a beehive, which is often two to four times larger. The nests had a small entrance, which often faced south and which, in most cases, was located in the lower part of the nest.

Remarkably, the walls of the nest cavity were entirely covered with propolis, a mixture of plant resins collected by the bees. The height of the nest entrance in the reported trees varied considerably. Seeley realised that when you ask people where bees' nests are, they report nests that are highly visible and therefore show only part of the variation. He therefore repeated the study, but this time by asking the bees themselves for the information. To do this, he followed the bees as they returned to their nests. This is not an easy thing to do. Seeley wrote an entire book on the method of following bees back to their hives. He managed to find 21 nests this way. They were all at least four metres off the ground.

    He thinks there are several explanations for the results he found. It could be that the bees occupied relatively small nesting cavities, as larger cavities are very rare. It could also be that the bees had nests at a height of more than four metres because there are more cavities and the entrances to these cavities are more often found on the south side of the trunk. But it could also be that the bees prefer small nesting cavities of 40 to 50 litres and that they prefer a nest high up in the tree with a south-facing entrance. To find out, Seeley conducted experiments in which swarms of bees could choose between two hives that differed in size, height, entrance size or compass direction. Seeley worked throughout the winter to create 252 different nest boxes. In the spring, he placed combinations of two hives in different locations in a forest called Arnot Forest. The wild bees living there produced enough swarms for him to be able to determine the choice of nesting box for no fewer than 124 colonies over two seasons. The choice of bees matched his earlier results surprisingly well. Now Seeley knew with certainty which characteristics bees prefer for their nest cavities. In fact, they seem to prefer to occupy cavities of around 40 litres. They also seemed to prefer caveties with a small entrance at the bottom, and cavities with an entrance on the south side. They also chose high-placed cavities and avoided low-placed cavities. Colonies of bees that choose their own nest cavities therefore live very differently from bees in hives belonging to beekeepers. Beekeepers' hives are much larger than the bees' preferred hives, with a capacity varying between 60 and 120 litres. Thomas Seeley's wild bees never chose a 100-litre hive. In a similar study, Tom Rinderer found that bees never chose a hive of 80 litres or more. Beekeepers' hives also tended to be low to the ground, whereas wild bees avoided low hives. Wild bees also preferred hives with smaller openings than the usual hives.

    There is another important difference between a hive and a natural cavity in a tree. The walls of a tree cavity are much thicker than those of a hive and, thanks to this thickness, they insulate the nest cavity, so they have to burn less honey in winter than in a hive to maintain the temperature. Heat loss in a tree cavity is easily four to seven times less than in a hive. The thick walls of a tree cavity also mean that the bees have to spend less time ventilating and cooling the nest on hot summer days.

Then there is the observation from Seeley's first study that the walls of bees' nests in a natural cavity have a layer of propolis that surrounds the entire nest, except for the entrance. As beekeepers often open the hives for all sorts of reasons, this layer of propolis is constantly being torn away. Beekeepers also prefer to keep bees that put little propolis in the nest.

The preference of bees for the characteristics of nest cavities has arisen through a long process of evolution by natural selection. This preference is therefore functional. The energy management of a bee population is very important here. In a nesting cavity of around 40 litres, a population of bees can develop sufficiently to reproduce by producing swarms. A larger cavity means greater heat loss, which reduces the chances of them surviving the winter. Heating costs are also higher during the season if the hive is kept at 35.5°C, which is the optimum temperature for the growth of young bees. A nest entrance at the bottom of the cavity prevents the heat produced by the bees escaping with the rising hot air. A small nest opening is also important for limiting heat loss. In addition, a small nest opening is easier to defend against honey predators such as wasps, and predators such as Asian hornets. An elevated nest position also helps to combat predators. It reduces the risk of bears or badgers attacking the nest and of mice stealing the honey in winter. The propolis envelope has antibiotic properties and protects the bee population from fungal and bacterial infections. A south-facing opening allows the entrance to the nest to be warmed by the morning sun, enabling the bees to become active earlier. The characteristics of the hives in which beekeepers place swarms of bees are therefore very different from those preferred by the bees.

The bees studied by Seeley live in a climate with long, harsh winters. There, a hive can survive the winter with a reserve of 25 kilos of honey. The nest cavity must therefore provide the space needed to build up such a reserve. Bees in milder climates can make do with a smaller supply. Researchers found that bee swarms in Costa Rica and the southern United States chose nest cavities much smaller than 40 litres. They also found that bees of the Italian subspecies *A. m. ligustica* accepted smaller nest cavities than bees of the subspecies *A. m. carnica*, which come from a cool mountain climate. The bees therefore adapt locally to the climate in their choice of nesting cavity.

Bees placed in hives by beekeepers miss out on the advantages of nesting cavities chosen by the bees themselves. The features of beehives in beekeeping serve the beekeeper first and foremost : they make working with bees less arduous and offer space for large colonies and for storing lots of honey.

Recommended reading

Thomas D. Seeley, 2016. Following the Wild Bees The Craft and Science of Bee Hunting. Harvard University Press, Cambridge Massachusetts.

Thomas D Seeley 2010. Honeybee Democracy. Harvard University Press, Cambridge Massachusetts.

Derek Mitchell, 2016. Ratios of colony mass to thermal conductance of tree and man-made nest enclosures of *Apis mellifera*: implications for survival, clustering, humidity regulation and *Varroa destructor*. Int J Biometeorol 60:629–638. DOI 10.1007/s00484-015-1057-z

Chapter 14:
Honey bees in the wild

Dorothy Galton was born in 1901 into a strongly left-wing family and later joined the Communist Party herself. Although she did not complete her university education, she developed a strong scientific interest in all aspects of Russian history and culture. As a member of the Communist Party and an expert on Russia, she was suspected by the British security services of being a spy for the Russians. No proof of this has ever been found. What is certain is that she combined her interest in honey bees with her interest in Russia and wrote a book entitled « *Overview of a Thousand Years of Beekeeping in Russia* ». That's how we know that Leo Tolstoy was one of the first to use hives with removable frames during the few years he kept bees and wrote War and Peace. More importantly for us, she described how wild bees were exploited in Russian forests. Beekeepers sought out the nests of wild bees in hollow trees. They also created their own 40-60 litre cavities in the trunks of living trees. In all the cavities they exploited, they made a sealable opening in the side by sawing off a slice of the trunk. This slice was put back in place after inspection of the nest. We know that around 10% of these tree cavities were occupied by bees. This means that the number of cavities was not limiting for the number of bee nests. It is possible that the density of wild bees was somewhat lower due to the exploitation of nests, although beekeepers never harvested more than six kilos of honey per year from a hive. In any case, the density of bee nests in these vast forests was 0.5 per square kilometre. Other estimates of the density of wild bee colonies in Europe range from 0.11 to 3.2 nests per square kilometre. Thomas Seeley found a density of 1 nest per square kilometre in the Arnot forest and quotes another estimate of 2.7 colonies per square kilometre. In tropical and subtropical areas, the density can be higher. The variation reflects the food supply in different places. The low density of wild bees in the wild in Western Europe and the northern United States shows that the distance between nests is much greater than in an apiary. This distance plays a role in the risk of infection by parasites and pathogens.

Although little research has been carried out on wild bees in Europe, many researchers assume that they have been largely wiped out by the invasion of the varroa mite. However, it is likely that local populations survived the invasion. This is suggested by Thomas Seeley's study of wild bees in Arnot Forest.

As a student, he carried out an inventory of wild bee colonies (Chapter 13), well before the invasion of the varroa mite in the United States. He was then able to use these data to measure the effect of the varroa mite on the wild bee population. With this knowledge, he was able to establish that the density of wild bee colonies in Arnot Forest is now as high as it was before the arrival of the varroa mite. This does not mean that the varroa mite invasion has gone unnoticed by these bees. Research into the genetic make-up of bees shows major differences between pre- and post-invasion bees. This can be explained partly by the fact that many colonies destroyed by the mites were replaced by new immigrants, and partly by the fact that the bees adapted by natural selection on the survivors. Thomas Seeley then set out to find out why wild bees suffered less from the varroa mite invasion than beekeepers.

As we saw earlier, wild bees live in tree cavities that have a much smaller volume than a hive. This forces wild bees to swarm earlier and more often than their captive counterparts. The old queen who is about to leave the nest stops laying before swarming, and the young queen who takes over the nest has to make a nuptial flight before she can start laying. There is therefore a period of at least three weeks during which there is no brood in the nest. The queen who leaves with the swarm cannot start laying again until the combs have been built in the new nest. The varroa mite cannot reproduce during this period, so the number of mites increases less rapidly in nests that swarm. Seeley experimented with bees in hives and showed that colonies that swarmed were less affected by varroa mites. He also knew that the average distance between wild bee nests could be as much as one kilometre. In the same experiment, he compared varroa infestations in colonies grouped together in an apiary with colonies isolated by distance. Of the colonies that had swarmed, the group that had been in the apiary together still had a large number of mites at the end of the season. But the colonies that had swarmed and were isolated by distance had far fewer mites at the end of the season and could therefore survive the winter. Neighbouring colonies lost the initial advantage of swarming during the season, as drones and workers who stole honey from neighbouring hives infected with varroa mites brought mites from other colonies into the hive. The great distance between wild bee colonies and their frequent swarming behaviour mean that they are less affected by varroa mites than captive bees. As a result, the survival rate of wild bees is higher and natural selection in these colonies is more likely to develop resistance through evolution. This is exactly what happened to the bees in Arnot Forest. David Peck was able to show in 2015 that honey bees in Arnot Forest have become resistant to varroa mites, using both defensive grooming behaviour and varroa-sensitive hygienic behaviour to render the mites harmless.

In Europe too, there could still be wild populations of honey bees that have become resistant in this way. The problem is that Europe has fewer extensive forested areas than North America and the human population density is much higher, so the density of hives is also greater. The bees in Arnot Forest were able to become resistant because the population had little contact with bees kept by beekeepers. In Europe, such a situation is rare, so wild bees are more likely to come from swarms that have escaped from beekeeping, and they are more likely to mate with bred drones, which have few resistance genes.

Research on the bees of Arnot Forest shows that a population of wild bees can provide an important buffer against extinction and promote the evolution of resistance to disease and parasites.

Recommended reading

Thomas D. Seeley. The lives of bees. The untold story of the honey Bee in the Wild. Harvard University Press, Cambridge Massachusetts.

Thomas D. Seeley , David R. Tarpy , Sean R. Griffin, Angela Carcione , Deborah A. Delaney, 2015. A survivor population of wild colonies of European honey bees in the northeastern United States: investigating its genetic structure. Apidologie 46:654–666.

Wild honey bees in hollow ash tree

Russian tree nest

Chapter 15:
Honey bees: indigenous, wild, domesticated or simply kept?

The poet Fernando Pessoa published numerous works, not only under his own name, but also under 75 other names. He used these "heteronyms" to defend unpopular or rather extreme positions. A swarm of different personalities thus represented the poet Pessoa. While Pessoa himself chose to adopt different characters, the honey bees suffer the unattractive fate that people attribute different characters to them. These characters then take on a life of their own, so that not everyone means the same thing when they talk about honey bees.

For example, in reports about the negative effects of honey bees on bumblebees and solitary bees, the problem is framed as the threat that honey bees pose to wild bee species. This suggests that honey bees are not counted as wild bees. Are there any arguments for excluding the honey bee as a species from the "wild bee" group, even though it is a native species? Firstly, there is the distinction between captive and non-captive animals. This distinction is enshrined in European law. Captive animals are entitled to care. The keeper has a duty of care. Non-captive animals generally look after themselves and live in the wild or what is left of it. It is also possible to keep wild animals. Just look at all the exotic birds and fish that have been captured in faraway countries to spend their lives here as cage birds or aquarium fish.

We can see straight away that it is difficult to classify honey bees as a species in one of these two categories. There are non-captive honey bees that nest in hollow trees, chimneys and hollow walls, and captive honey bees that live in hives. Captive honey bees are also largely self-sufficient and live largely in the wild, making the distinction even more difficult.

Are there any other criteria for concluding that honey bees are not "wild pollinators"? Many species of animal have been kept by humans for a very long time. Kept populations of these animals often differ strikingly from their wild counterparts in terms of shape, size or colour. They have been selected for characteristics that people found useful or beautiful and, as a result, these populations differ from their non-captive counterparts in terms of genetic characteristics. A captive population of a species that also differs genetically from its non-captive counterparts is said to be domesticated. The genetic differences between domesticated animals and their wild counterparts can be enormous, and domesticated animals are often no longer capable of living in the wild. Bellier rabbits and Chihuahua dogs, for example, would not survive long in the wild. Genetic differences can also be very small; a colour mutant dwarf hamster may differ from its wild ancestor by only one allele. The question is whether we consider such a small difference to be sufficient to classify an animal as domesticated. This brings us to the question of whether domesticated populations of honey bees exist. To answer this question, we need to study the extent to which bee breeds selected by beekeepers can be considered domesticated. These bees do not differ in shape, size or colour from wild bees. They function exactly like non-captive congeners when it comes to searching for and gathering food, and their mating behaviour does not differ from that of non-captive congeners either. An escaped swarm can survive in the wild just like a colony of non-captive bees. There are therefore few arguments for describing honey bees as domesticated. Nor can this qualification be used to separate honey bees from "wild pollinators".

Honey bees have been exploited by humans since ancient times, but because of their particular mating biology (chapter 6), they have hardly been domesticated. In this respect, they are very similar to mussels and oysters : wild species that are also farmed, while farmed and non-farmed animals are genetically the same population. In the case of oysters, this applies to the native flat oyster, but also to the exotic Japanese oyster, which has become wild here. The honey bee is a native species. The *A. m. mellifera* subspecies has at least lived in Western Europe since the last Ice Age and possbly as long as 1 million years.

A final criterion could be the fact that many honey bees are exotic or hybrids with an exotic parent, such as the Buckfast bee bred by Brother Adam. It could in fact be said that these exotics have no place here and therefore cannot belong to the wild bees. It follows from the above that the designation "wild" or "non-wild" cannot be applied to an entire species, but only to sub-populations, which are most often made up of captive animals, but sometimes of domesticated captive animals, of exotic or non-exotic origin.

Another problem with honey bees is the importance of subspecies. I sometimes meet people who think that protecting the black bee is nonsense. Their argument is that the black bee is only a subspecies, not a species, and therefore does not qualify for protected species status. So these people have no objection to crossing subspecies. After all, they all belong to the same species. It is therefore worth remembering that species are units defined by humans. The subspecies of the honey bee are cross-breedable and can produce fertile offspring. But without human intervention, this would have been rare, as most subspecies would never have met. They are isolated from each other by mountains, seas or deserts. More importantly, they each have an independent evolutionary history of around 1 million years and have adapted to their own environment over this period. The fact that they are crossbreedable is only a relative fact. Did you know that jackals and wolves are cross-breedable and that the hybrids are fully fertile? There are many other examples of fertile crosses between different species. Being cross-breedable is therefore no reason to treat our native bees lightly and wipe out 1 million years of evolution!

Recommended reading

Eva Crane, 1983. The Archeology of Beekeeping. Duckworth, London

Julia Gabryś, Barbara Kij, Joanna Kochan, Monika Bugno-Poniewierska, 2021. Interspecific hybrids of animals in nature, breeding and science – a review. Annals of Animal Science 21: 403–415

Chapter 16:
Enemies from the Far East (1): The varroa mite

Just south of Vladivostok, sandwiched between China and North Korea, lies a narrow strip of Russia officially known as the 'Maritime Territory', but known locally and in the world of beekeepers as Primorsky. Gold was found there, the sea was rich in fish and there were vast forests. According to Eva Crane, around 1850, Russians immigrated from the West to make a new life here. The Oriental honey bee, *Apis cerana*, which is native in the Primorsky region, lives wild in the forests. Immigrant farmers have tried to keep these bees. This proved difficult, as the bees often absconded their hive, leading to the loss of many colonies. In 1904, the Trans-Siberian Express came into service and a new wave of immigration followed; between 1906 and 1914, four million farmers arrived from Western Russia and Ukraine. They took their Western bees with them on the train. These immigrants were unaware that a parasitic mite called *Varroa destructor* lived in the native honeybee colonies of the East. Over time, the varroa mites moved to western honey bees. These bees had a reputation for being excellent honey producers. However, the fact that this reputation was due to the vast forests of lime trees that provided an inexhaustible source of nectar for the bees, and not to the quality of the bees themselves, only became apparent when the Russians returned to Moscow with their bees around 1950, and production was disappointing. Unfortunately, with these bees, varroa mites arrived in the West. They are now widespread throughout the world and became recently established in Australia. They are a major pest, killing many bee colonies every year.

  The life cycle of the varroa mite is completely adapted to that of the honey bee. An adult female attaches herself to a worker bee and feeds on her body fluids. When the worker is looking after a bee larva about to pupate, the mite quickly leaves the worker and crawls under the larva to the bottom of the cell. There, it waits until the larva has woven a cocoon and is about to transform into a chrysalis. Meanwhile, the workers have sealed the cell with a wax cover. The mite then lays a total of three or four eggs. The first egg gives birth to a son, who then mates with his sisters, who hatch from the eggs laid afterwards. The mother pierces a hole in the abdomen of the developing bee pupa and the young mites and their mother feed through this hole. The parasitized bee does not die, but becomes a somewhat weakened worker. When this worker leaves the cell, the varroa mother and her fertilised daughters also leave the cell, in search of new reproductive opportunities. The son stays behind and dies. In winter, when the bees are not reproducing, the mites settle on the overwintering bees and feed on their body fluids. If the mites are numerous, this can lead to the death of the entire colony. Varroa mites also appear to transmit a number of viral diseases. In European and North American bee colonies, the combination of untreated mite infestation and viral diseases ultimately and irrevocably leads to the death of the colony.

The Oriental honey bee, the original host of varroa mites, has developed defence mechanisms during the joint evolution of the parasite and the host. As a result, mite numbers remain low and colonies suffer little damage from the parasite. Western bees do not have this common history, but it has been shown that the genetic basis of some of the defence mechanisms is sometimes present.

In 1936, long before varroa mites were known, O.W. Park discovered that bees resistant to the bacterium responsible for "American foulbrood" eliminated infected larvae from their cells. He called this activity hygienic behaviour. It later transpired that these hygienic bees also eliminated larvae infected with other diseases and dead larvae. Already in 1991, Otto Boecking and Wilhelm Drescher discovered that hygienic bees sometimes also eliminated varroa-infected cells. In 1999 they published evidence that bees responding to varroa-infested cells with hygienic behaviour are genetically different from those removing only dead bee larvae en called the behaviour « varroa specific hygiene ». At the same time, John Harbo and his colleagues worked on selecting varroa-resistant bees. To find out how the bees managed to survive infection by the varroa mite, they opened cells containing pupae in the combs of these bees and compared them with non-resistant bees. In the survivors, they found more cells in which there was a female varroa mite, but no offspring. In non-resistant bees, they generally found mites with offspring. They concluded that the resistant bees inhibited varroa mite reproduction. In a telephone conversation with John Harbo, Marla Spivak suggested that hygienic behaviour might play a role after all. So John Harbo decided to carry out further experiments. And yes, the bees mainly eliminated the varroa mites with offspring, leaving the females without offspring. So the bees did not suppress mite reproduction, but prevented it by emptying the cells containing mite families. But how do the bees recognise the cells containing mite families? Fanny Mondet suspected that the parasitized cells gave off an odour, and then showed that this was indeed the case. Is it the mites that have produced the smell, or does the smell come from the young bee that has been attacked? The mites should try not to betray their presence, but the attacked larva would do well to use an olfactory signal to warn its sisters that parasites are developing in its cell. Fanny has been able to show that the smell does indeed come from the bee. This reveals a new and tragic aspect of communication in a bee colony: young bees invite their adult sisters to come and kill them in the interests of the colony's survival.

Otto Boecking called the defence behaviour varroa-specific hygiene. John Harbo called the same behaviour varroa-sensitive hygiene. Fortunately, both names have the same abbreviation: VSH. John Harbo and Jeffrey Harris have shown that VSH is a heritable property that can be selected for in order to breed varroasis-resistant bees. Since VSH is a heritable trait and varroa mite is a major cause of mortality, one would expect natural selection to rapidly lead to varroa resistance.

The behaviour of eliminating varroa-infected worker pupae is strongly present in the Oriental honey bee. As a result, the varroa mite cannot reproduce on workers and must be confined to drones. The Oriental honeybee has another defence strategy against the varroa mite. If an adult mite clings to a worker, she tries to get rid of it by violent grooming movements with her legs. If the mite is in a place she can't reach, she invites other workers to help her clean it up. In this way, she often succeeds in removing the mite from the bee. The bees try to catch the mites that wander onto the combs with their jaws and bite them to death. Together, these two defence methods ensure that mite numbers remain low in Oriental honeybee colonies.

Could Western honey bees also get rid of varroa mites by grooming them off or allowing other workers to remove them?

Western honey bees also show grooming behaviour regularly, as do other bees, usually to remove dust and dirt and sometimes to get rid of another parasite, the tracheal mite. However, varroa mites are not easy to get rid of. Mites are very fast and can attach themselves very well to bees. Only a rapid response with vigorous brushing movements works, and even then the mite still escapes regularly. Most western bees react too late and too little when they are accosted by a varroa mite and are therefore not very effective at eliminating it. A small proportion of bees are able to get rid of a mite by grooming behaviour. The grooming behaviour of bees in a hive is difficult to observe, and knowledge of the importance of grooming as a defence against varroa mites comes mainly from indirect measures, such as counting dead and broken mites that have fallen to the floor of the hive. The data seem to indicate that this behaviour is heritable, so we might expect natural selection to make it more common too. Greg Hunt and his colleagues have in fact been able to select bees that are less affected by varroa mites because of this behaviour: mite-biting bees from Indiana.

The next chapter tells how natural selection has enabled Western honeybee populations to become resistant to the varroa mite in South Africa and South America. This chapter also explains why large numbers of bee colonies in Europe and North America still die each year from varroa infection. The paradoxical answer is revealed here : beekeepers' practices work against natural selection for resistance.

Recommended reading

Jacques J. M. van Alphen and Bart Jan Fernhout, 2020. Natural selection, selective breeding, and the evolution of resistance of honey bees (Apis mellifera) against varroa. Zoological Letters. https://doi.org/10.1186/s40851-020-00158-4

O Boecking, K.Bienefeld 2 and W. Drescher, 2000. Heritability of the varroa-specific hygienic behaviour in honey bees (Hymenoptera: Apidae). Journal of Animal Breeding Genetics. 117: 417-424.

J.R. Harbo and J.W. Harris, 1999: Heritability in honey bees (Hymenoptera: Apidae) of characteris- tics associated with resistance to *Varroa jacobsoni* (Mesostigmata: Varroidae). J. Econ. Entomol. 92: 261±265.

Nuria Morfin, Krispn Given, Mathew Evans, Ernesto Guzman-Novoa and Greg J. Hunt, 2020. Grooming behavior and gene expression of the Indiana "mite-biter" honey bee stock. Apidologie (2020) 51:267–275

Chapter 17:
The evolution of varroa resistance through natural selection

In South Africa, varroa mites were first discovered in Stellenbosch in 1997, probably imported with a commercial shipment of bees and queens. Mike Allsopp has followed the development of this parasite and thanks to him we now have a detailed picture of its beginnings. Two subspecies of honey bees are present in South Africa: the Cape honey bee (*A. m. capensis*), along the southwestern and southern coasts of South Africa, and the savannah honey bee (*A. m. scutellata*) in the rest of South Africa. The mite spread rapidly and was found throughout the country within five years. At its peak, a colony could contain an average of 10,000 to 17,000 mites, and sometimes as many as 30,000 to 50,000. Apparently, the mite was initially able to reproduce very efficiently in colonies of both subspecies. At the height of the infestation, 30% of colonies succumbed to the mite infection, but the majority survived. The tolerance of Cape and Savannah bees to very high infection rates is probably due to the absence of the harmful virus epidemics that always accompany varroa infection in Europe. After the peak of infection, the density of mites gradually decreased and bees in the Cape became resistant three to five years after the arrival of the varroa mite, while bees in the Savannah became resistant after six to seven years. Since then, the varroa mite has no longer been a problem in South Africa: according to Mike Allsopp, "the varroa mite is now just a random presence". The natural selection of varroa-sensitive hygiene and grooming behaviour had led to this resistance. In Europe and the United States, the varroa mite invasion has led to a very different situation. Almost all the bee colonies that have been infected have succumbed, unless the mites were controlled chemically. This has led to a situation where every beekeeper treats his colonies at least once a year with chemical agents against varroa mites. Initially, beekeepers mainly used acaricides, but the mites quickly became resistant to these agents, which also left residues in the honey and wax. Today, natural acids (formic acid and oxalic acid) are the main agents used to combat varroa mites, along with vegetable oils such as thymol. Chemical control of varroa hinders the natural selection for resistance, so that bees without hygienic behaviour can also survive. In addition, a number of standard procedures are used in beekeeping, which also work against the natural selection of resistance. For example, beekeepers with pedigree bees replace the queens in their colonies every year. They also send virgin queens to a mating station, where they then mate - until now - with non-resistant drones. The drone brood is often removed from the hives to reduce varroa infection. Chemical control and the beekeeping measures mentioned above prevent natural selection from developing resistance. This is why, after forty years, the varroa mite problem is still as important as ever.

The high mortality of bee colonies following the introduction of varroa mites has made chemical control necessary. The question is why this high mortality occurred in Europe and the United States, but not in South Africa. The reason why European bees seem so vulnerable to varroa mites is that alleles for hygiene and cleaning behaviour against varroa-susceptible mites are very rare (chapter 9). For years, beekeepers have selected queens that produce beautiful, regular brood. If, in a comb of worker pupae, there were open cells here and there, this situation was called "hazy brood". The idea was that the queen had skipped these cells during laying and that this was undesirable. In the meantime, we know that there is another explanation for the holes in the brood: workers behaving hygienically remove diseased or dead pupae. So it seems that by selecting for regular brood, beekeepers have selected against hygienic behaviour, and so the alleles for this behaviour have become rarer. Another reason why European bees are so vulnerable to varroa mites is their susceptibility to viruses spread by the mite, in particular the deformed wing virus. In heritable variation for resistance to these viruses has been found in honey bees in Europe, but alleles for this resistance are also very rare.

This brings us to the question of why varroa and virus resistance genes are so much rarer in Europe and North America than in South Africa. A key difference in the way bees are kept between South Africa, on the one hand, and North America and Europe, on the other, is that honey bees in South Africa can mate freely. There are many colonies of wild bees throughout South Africa. Captive bees sometimes abscond the hives they inhabit and return to the wild, while beekeepers capture swarms of wild bees and place them in hives. The young queens mate with the drones in the drone congregations, and the bees form a large mixed population. In contrast, the queens of European and North American bees mate with selected drones in isolated mating stations or are artificially inseminated. The queens often come from a queen farm, which means that several queens have the same mother. In addition, many beekeeping measures aim to prevent the annual reproduction of bee colonies.

European and North American beekeepers have thus deprived their bees of the most important weapon in the fight against diseases and parasites, namely that queens can mate with the best drones in a large area and with many drones, which, moreover, differ as much as possible in terms of genetic characteristics so that the extremely high frequency of recombination can also be effective (Chapter 8). The forty-year fiasco of the varroa mite epidemic shows that there is something fundamentally wrong with the way beekeepers have bred bees to be highly productive and less aggressive. Breeders have failed to take into account the high susceptibility of honey bees to bacterial and viral infections, due to the numerous contacts between honey bees within a colony and the temperature of the brood nest. Breeders have also underestimated the importance of all aspects of bee mating biology and why mechanisms to maximise heritable variation are so important in defending against bacteria and viruses.

However, it is possible to make European pedigree bees resistant to varroa mites and varroa-transmitted viruses, because the genetic traits needed for this resistance are present, although there are too few of them to be effective at present. Selection, either by nature or by the beekeeper, could increase the percentage of resistance genes. If chemical control of varroa mites and certain beekeeping practices go against natural selection, natural selection could be expected to work if chemical control of varroa mites and the above-mentioned beekeeping interventions were not used. Some European bee researchers have tried to breed varroa-resistant bees in this way. All but one of these attempts failed. The exception is the experiment by Ingemar Fries. He worked at Uppsala, at the Swedish University of Agriculture, and set up a vast experiment to study whether natural selection could lead to varroa resistance in bees. He placed 150 colonies of bees on a peninsula south of the island of Gotland in the Baltic Sea. The colonies were thus isolated from the bee populations on the Swedish mainland. The isolation of these experimental colonies is the reason why the experiment was more or less successful. All the other attempts involved colonies that had not been treated with pesticides, but which were not spatially separated from colonies that were not part of the experiment. In this situation, although severe natural selection occurs in the experimental colonies, the non-resistant colonies quickly succumb and the colonies that happen to have a higher percentage of resistance alleles can survive. But the young queens from the surviving colonies then mate mainly with drones from colonies outside the experiment, and these have very few resistance genes, which again dilutes the resistance genes in the new population. Since honey bees mate with drones from a very wide area (see chapter 7), the higher percentage of resistance genes obtained by selection is also lost again and again in subsequent generations The surviving colonies should therefore die out quickly, if the researcher did not repeatedly rear a large number of young queens from those colonies. These in turn mate with non-resistant drones in the area, so these selection experiments are a constant stopgap. Examples of these failed attempts come from France (Le Mans and Avignon) and southern Norway. Similar experiments have also been carried out in the Netherlands, by Tjeerd Blacquière and his colleagues over the period 2006-2016. Although the authors claim that the experiments led to varroa-resistant bees, no results have ever ever been published to demonstrate this.

Ingemar Fries avoided the problem of resistance alleles escaping. His experimental bees were sufficiently isolated by the narrow land bridge between the peninsula and the main island. This meant that the young queens had to mate with drones from the surviving colonies. The colonies were not treated for varroa mites. Swarms were added to the population as new colonies. After four years, 38 new colonies were formed by swarming, but varroa mortality meant that only 13 colonies out of 188 survived. The high mortality resulted in a huge loss of genetic variation, and the surviving colonies showed all the characteristics of inbreeding. These colonies were then multiplied using beekeeping methods. They proved capable of surviving without varroa control, but remained small. The colonies proved resistant to the viruses to which bee colonies usually succumb after varroa infection. It appears that the genes for varroa-sensitive hygiene and cleaning behaviour were not present in the small number of surviving colonies. They were therefore not varroa-resistant, but tolerant, as they did not succumb to the viruses. Because of inbreeding, they produce little brood, which means that the varroa mite can only reproduce to a limited extent. What's more, the honey bee season in Sweden is short, which also limits the opportunities for varroa mites to reproduce. Ingemar Fries' experiment therefore produced honey bees capable of surviving infection by the varroa mite, but these bees could not serve as the basis for a solution to the varroa mite problem in Europe and the United States.

When comparing the evolution of the varroa mite epidemic in South Africa and Europe, an important difference in the structure of honey bee populations is crucial. In Western Europe, honey bee queens mate with drones from selected colonies in an isolated mating station. Consequently, in Western Europe, in pedegree honey bees, there are no populations of honey bees in which natural selection can operate. In South Africa, not only are there numerous colonies of wild bees, but the honey bees also mate freely with each other and with the wild population. Since African bees also migrate frequently, there is an unlimited exchange of genes over great distances. The free exchange of genes within a very large population, the extreme polyandry of queens and the very high frequency of recombination (chapter 8) are the bees' weapons in the arms race against viruses, bacteria and parasites such as the varroa mite (chapter 15).

By comparing the situation in Europe with that in South Africa, we can also conclude that natural selection for resistance can only work when the percentage of resistance genes in the bee population has become so high that a high proportion of colonies can survive without chemical control of the varroa mite. Until this happens, most beekeepers will not want to stop using chemical control. This impasse could be broken by artificial selection aimed at increasing the percentage of resistance genes. This is the subject of chapter 19.

Recommended reading

Michael Allsopp. Analysis of *Varroa destructor* infestation of Southern African honeybee populations. MSC Thesis Nat Agric Sci Univ Pretoria. 2006; (June): 285pp.

Ingemar Fries, Henrik Hansen, Anton Imdorf, Peter Rosenkranz, 2003. Swarming in honey bees (*Apis mellifera*) and *Varroa destructor* population development in Sweden. Apidologie 34: 389–397

Ingemar Fries, Anton Imdorf, Peter Rosenkranz, 2006. Survival of mite infested (*Varroa destructor*) honey bee (*Apis mellifera*) colonies in a Nordic climate.
Apidologie 37 (2006) 564–570

Chapter 18:
Varroa mites in South America

On 15 September 2018, Warwick Estevam Kerr passed away. He was the son of Scottish immigrants who came to Brazil via the United States. He was a respected scientist with an impressive career, best known for his studies into the genetics of stingless bees native to Brazil. Warwick E. Kerr was a pupil of Theodosius Dobzhansky and a leading genetics researcher in Brazil. But when he died, the newspapers ran the headline: "The man who created killer bees is dead", as if he had been Victor Frankenstein's successor. But what had happened?

Kerr had noticed that bees imported from Europe were poorly adapted to Brazil's tropical climate. He decided to import honey bees of the subspecies *scutellata* from Tanzania and South Africa, with the intention of using them to improve the Brazilian bees. In 1956, 51 colonies were imported and placed in hives fitted with a queen excluder so that the drones and queens could not leave the hives and swarm, but due to a mistake by one of the assistants, the excluders were removed and 26 colonies disappeared into the Brazilian wild. There, they felt perfectly at home and multiplied rapidly. Because of the great migratory urge characteristic of African bees, every year they could be found 300 kilometres further away from where they had started. Today, they can be found everywhere in South America where the climate permits. They have also colonised the southern United States via Central America. The first generation of hybrids of these African bees with the European bees present in South America proved to be extremely aggressive. They defended their nests by attacking suspected intruders en masse. Some deaths received considerable press attention, although the number of people who died as a result of attacks by Africanised bees was lower than the number of deaths caused by native scorpions. Later generations, containing more African alleles, proved to be slightly less aggressive, but still much more aggressive than European bees. The gradual reduction in human deaths is thought to be due to a number of factors, including increased public awareness of the dangers of approaching bee nests and hives, the relocation of apiaries away from densely populated areas and the selection of less aggressive bee colonies by beekeepers. The bees introduced by Kerr have improved honey production and the pollination of economically important crops in Brazil.

Warwick Estevam Kerr was a critic of the dictatorial military regime in

Brazil at the time. He was arrested twice, but because of his great international reputation, they couldn't get rid of him. So the regime tried to destroy his reputation by portraying him as the creator of killer bees. The international press was quick to pick up on this information. In South America, Africanised bees have formed wild populations in tropical forests, where they are now very numerous and have become the main pollinators. In 1979, more than 20 years after the arrival of Africanised bees, varroa mites were found in Brazil. Unlike in Europe, the varroa mite invasion was not accompanied by large-scale mortality of bee colonies. Chemical control of the mites was therefore not necessary. It seems that, because of the high density of wild bees, natural selection rapidly led to resistance to the mites. Bees kept by beekeepers largely mate freely with their wild counterparts. This is one of the reasons why natural selection has been able to do its job. Colony losses of Africanised bees due to varroa mites have not been reported, and any negative effects of varroa mites on honey production appear to be negligible. This result is surprising because, unlike South Africa, the varroa-associated deformed wing virus is widespread in South America and bees are not resistant to it. In one study, the increase in the virus was lower in Africanised bees than in European bees, which could be a sign of natural selection against the virus. Africanised bees are resistant to varroa because they behave more hygienically than European bees. They are also more efficient than European bees at eliminating mites from their bodies through cleaning behaviour. The resistance of South American honey bees to the varroa mite is perhaps Warwick Estevam Kerr's most important contribution to South American beekeepers.

Recommended reading

Mark L. Winston, 1992. The Biology and management of Africanized honey bees. Annual Review of Entomology 37: 173–193.

M Spivak, DJC Fletcher, MD Breed, 2019. The "African" Honeybee. Routledge New York.

Stephen J. Martin and Luis M. Medina, 2004. Africanized honey bees have unique tolerance to *varroa* mites. Trends in Parasitology. DOI: https:/doi.org/10.1016/j.pt.2004.01.001

Chapter 19:
Selection for resistance to the varroa mite

One of the fascinating aspects of beekeeping is the inventive way in which beekeepers manipulate the biology of bees to maximise production. The mating biology of honey bees is a stumbling block in bee breeding because, in natural mating, beekeepers have no control over which drones a queen mates with. One solution is to take virgin queens to an isolated drone population, usually on an island, and let them carry out their mating flights there. But even if the queens mate in this way, the problem remains that the drones with which they mate do not all have the same characteristics and we do not know which drones they have mated with. For research into the inheritance of characteristics, it is desirable to have queens that mate with a single drone. The only way to be sure is to artificially inseminate the queen. Over the course of the twentieth century, a number of researchers worked on developing a method for doing this reliably. Today, the necessary instruments can be ordered via the Internet. Drone sperm can be collected by squeezing the abdomen of a sexually mature drone. The endophallus then expands and the sperm can be aspirated using a pipette. To inseminate a virgin queen, she must first be stunned with CO_2. She can then be placed in a tube so that her abdomen protrudes. She is then fixed in the insemination device. Two hooks are used to hold her abdomen open and push the tip of the pipette containing the sperm into the fallopian tube using a sledge. This operation requires experience and skill, as a partition in the fallopian tube must first be opened.

You might think that a queen inseminated in this way would go on to make mating flights, but this is not the case. Beekeepers have discovered that after a second carbon dioxide stun, the queen has lost her sexual desire and no longer goes on her honeymoon.

Chapter 8 tells us that bee colonies in which the queen has mated with many drones are more resistant to disease and gather food more efficiently than colonies in which the queen has mated with fewer drones. Artificial insemination with the sperm of a single drone therefore results in small, relatively weak colonies. However, it is possible to keep such colonies for at least two seasons. This is long enough to study the rate at which the varroa mite can multiply in these colonies.

In the 1990s, bee researcher John Harbo worked in the bee laboratory of the USDA, the research arm of the US Department of Agriculture, in Baton Rouge, Louisiana. Together with Jeffrey Harris and Robert Danka, they used the method of artificial insemination with sperm from a single male to discover the genetics of honey bees' resistance to varroa mites (chapter 15). The resistant colonies obtained from these experiments were not suitable for commercial use because of inbreeding, but the USDA considered research to be its main task. A possible solution to the varroa mite problem had been found, but much remained to be done to ensure that resistant bees were widely available to beekeepers.

Dutchman Bart Jan Fernhout has been an enthusiastic beekeeper since childhood. From his own experience, he knew how varroa mites can destroy bee colonies, but also that chemical control of these mites is not without risk. He also understood that selection for resistance at colony level cannot work in an open-mating population if the frequency of resistance genes in the population is very low. This intuition was not shared by many European bee researchers, who had been using this selection method for years without success. He had read John Harbo's publications and realised that the method of artificial insemination with the sperm of a single drone could not only be used as a research tool on genetics, but might also accelerate the selection of resistance. He decided to set up the foundation *Arista* with the aim of solving the varroa problem. They work as follows: the selection begins in spring with the rearing of queens. These young queens are then artificially inseminated and placed in small boxes with enough workers to start a colony. After just over a month, all the workers in such a colony will be daughters of the young queen. Beekeepers then take varroa mite females from untreated, varroa-susceptible colonies and infect the young colonies with a standard number of varroa mite females. These mites then infect the brood and attempt to reproduce. Beekeepers wait until the brood has formed young pupae in wax-sealed cells. They then open as many cells as necessary to get an idea of the degree of infection and the reproductive success of the varroa mites. In this way, they can determine whether a population has no, few or many resistance alleles. The best colonies are kept until the following season. Sometimes they breed queens from these colonies again in the same season, to make sure they have a better starting point for next year's selection, because queens always die during the winter and a lot of work could be lost if no extra queens had been reared from a promising colony. If they have obtained totally resistant colonies in this way, they will breed extra queens which they mate naturally with a large number of drones. The drones born in such a colony are totally resistant, as is their mother, because drones have no father. The drones from such a colony can then be used to mate naturally at an isolated mating station with young resistant queens from other resistant colonies. However, the single drone mating on which the method is based results in the quick loss of many rare resistance alleles and valuable genetic variation further decreases under the effect of selection (see chapter 9). Single drone insemination is a tool for honey bee geneticists and is not meant to be used in selection programmes. Only when beekeepers will agree to stop using chemical treatments against varroa mites, and allow natural selection for greater resistance to take effect, as it has in South Africa and South America, honey bees become resistant.

Recommended reading

John R. Harbo, 1999. The value of single drone insemination in selective breeding of honey bees. Apiculture for the 21 Century. Chapter 1. Wicwas press.

Chapter 20:
Enemies from the Far East (2): The Asian Hornet

In 2004, a ship arrived in Bordeaux with a cargo of Chinese earthenware jugs. One of the jars contained a nest of wasps, which had come as a stowaway. In 2005, more than 100 kilometres away, in the village of Nérac, an unknown wasp was found on a persimmon fruit. The wasp was identified as *Vespa velutina*, the Asian hornet. In 2006, three were found in the village of Villeton. The Asian hornet had unfortunately become established in Europe and soon appeared to be advancing northwards at a rate of around eighty kilometres a year. Not only does this hornet have an aggressive temperament, which makes it dangerous for people who get too close to its nest, but it also prefers to eat honey bees and can wipe out a bee colony in a short space of time. The Asian hornet is now widespread throughout France and has invaded Spain, Belgium, Germany and The Netherlands and the British Isles. The European hornet, *Vespa cabro*, which, with its larger body and shiny yellow-black markings, is more impressive than the Asian hornet, is not a specialist in capturing bees. As a result, most honey bees in Western Europe have never been exposed to such an enemy. Because of the absence of a common evolutionary pattern, but also because beekeepers have selected non-aggressive bees, the defence of European bees against this invasive exotic species has not been very effective.

The eastern honey bee, *Apis cerana*, has developed defence mechanisms against the Asian hornet in the course of their shared evolution. In East Asia, the giant hornet, *Vespa mandarina*, attacks bee colonies in addition to the Asian hornet. We saw earlier that honey bees emit a high-pitched buzzing sound during the shudder dance, which is a stop signal. This stop signal is also used by the Oriental honey bee when hornets are spotted near the nest. This prevents the bees from recruiting other bees and the bees stay in the nest. Oriental honey bees even have two different stop signals, one for the Asian hornet and one for the giant hornet. When a hornet enters the nest, the bees form a ball around the intruder and begin to vibrate their flight muscles. This causes the temperature inside the ball to rise to 45°C and the concentration of carbon dioxide to increase considerably. This kills the intruder, who is then unable to recruit other mates for a joint attack on the hive.

When the bees detect an approaching hornet in front of the hive entrance, they vibrate their abdomen. This is a signal to the approaching hornet: we see you! There is then the implicit threat that the guardians, aided by the other bees who come out of the nest to defend themselves, will try to trap the hornet in the ball they are forming. The enemy then often chooses to try his luck elsewhere.

In the Middle East and south-east Europe, there is another species of hornet that attacks bees' nests, the Oriental hornet, *Vespa orientalis*. This species is also found in south Eastern Europe. As a result, the Cypriot honey bee has evolved with this enemy. The Cypriot bee has two very different defence strategies. One strategy is to withdraw inside the nest. Bees that do this barricade the entrance to the nest with a wall of propolis, in which they leave only a limited number of small holes through which bees can enter and leave. These holes are too small for the hornets to pass through. Beekeepers in Western Europe who select bees that produce little propolis, because it is difficult to inspect the hives, make their bees vulnerable to hornets. Bees selected in this way no longer reduce the entrance to their hive. Native black bees sometimes reduce the size of the nest entrance in autumn, to protect themselves from the cold.

The Cypriot bees' other strategy is to attack. Colonies that use this strategy leave the nest opening free. If a hornet attack is imminent, part of the hive flows out and lands on the flight board or the wall of the hive. The bees make synchronised movements with their abdomen, which travel through the crowd of bees like a wave and produce a powerful hissing sound. The meaning of this expression is probably, as in the case of the Oriental bee, "We can see you". If the hornet approaches, it is trapped in a ball of bees and suffocates, as in the case of the Oriental honey bee.

Karine Monceau studied how black bees in the Bordeaux region react to attacks by Asian hornets. In this area, there was an average of twelve hornet nests per square kilometre. She found that the threat of hornets caused the bees to spend less time foraging, and that the number of worker bees declined over the course of the season as a result of predation. The two colonies that were constantly observed with video cameras had too few workers to survive the winter at the end of the season. However, Monceau also found that the bees behaved defensively. They run out of the nest when attacked and sometimes form a ball around the hornet that is attacking them. Although, in this case, the behaviour was not sufficient to limit predation pressure, her research shows that black bees are capable of reacting to an attack by the Asian hornet, and that selection for this trait could lead to effective resilience in bees against this new enemy.

Meanwhile, national governments in the European Union are trying to limit the spread of the Asian hornet as much as possible. Found nests are being removed and traps have been developed to catch young queens in the spring and workers later in the season. The problem is that these traps also catch other wasp species. Unfortunately, it is no longer possible to completely eradicate the Asian hornet. Bees and people will have to learn to live with this bee killer.

Recommended reading

M. Arca, F. Mougel, Thomas Guillemaud, S. Dupas, Q. Rome, et al. 2015. Reconstructing the invasion and the demographic history of the yellow-legged hornet, *Vespa velutina*, in Europe. Biological Invasions,17: 2357-2371. ff10.1007/s10530-015-0880-9ff. ffhal-02370176f

Karine Monceau, Olivier Bonnard and Denis Thiéry, 2013. *Vespa velutina*, a new invasive predator of honey bees in Europe. Journal of Pest Science.87: 1-16.

Karine Monceau, Nevile Maher, Olivier Bonnard, Denis Thiéry, 2013. Predation pressure dynamics study of the recently introduced honeybee killer *Vespa velutina*: learning from the enemy. Apidologie, 44: 209-221. ff10.1007/s13592-012-0172-7ff. ffhal-01201288f

Chapter 21:
Self medication in honey bees.

Plants defend themselves against herbivores and disease by producing a large number of substances which do not play a role in the plant's metabolism, but which offer it protection. These substances are known as secondary plant substances. We know them well, as they give flavour to tea, coffee, wine and beer, and aroma to aromatic herbs such as garlic, bay leaves and thyme. Plants that depend on insects for pollination produce nectar. Nectar is an aqueous solution of sugars, making it an ideal breeding ground for yeast and bacteria. To prevent nectar from fermenting quickly and losing its appeal to pollinators, plants add antimicrobial substances. The bees that gather the nectar therefore return home not only with a precious cargo of sugars, but also with a pharmacy full of secondary plant substances. Different types of plant produce different secondary plant substances, and the effects on bee health depend on the plant species from which the nectar comes. It has been shown that honey bees selectively choose between different types of nectar depending on their state of health. This can be important because the protective effect of secondary plant substances contained in the honey of a particular plant species depends on the disease. For example, black locust honey has a strong inhibitory effect on American foulbrood, while sunflower honey has a stronger inhibitory effect on European foulbrood. Bees can taste low concentrations of secondary compounds in nectar and therefore have a mechanism for selectively choosing nectars from different plant species. Pollen also contains secondary plant compounds, which have an antioxidant effect and protect against infection. These are species-specific phenols. Bees store pollen in the form of "bee bread". Bee bread is made up of pollen mixed with bee saliva and covered with a drop of honey. Stored in this way, it undergoes a maturing process consisting of fermentation by lactic acid bacteria and a number of Bacillus species present in bee saliva. These bacteria are able to develop in a phenolic environment and produce substances that inhibit the growth of other bacteria and moulds. Fermentation increases the concentration of phenols and, consequently, the antibacterial and antifungal effect of pollen.

Trees and shrubs often protect their young leaves and buds with a layer of resin, which also consists of secondary plant substances. Bees collect this resin to make propolis, a mixture of plant resins and beeswax. Wild colonies in tree cavities and domestic colonies in commercial hives use propolis to cover holes and crevices in the nest and to make the entrance to the hive smaller. This is also where the word "propolis" comes from: pro means "for" and polis "city". Wild bees use propolis, among other things, to cover the inside of their nest cavity in a hollow tree with a protective layer. In this way, propolis prevents draughts, damp and leaks, and keeps intruders out. Propolis is also used by man to prevent infections, reduce them or heal them after an infection. Honey bees also use propolis to protect their hives from parasites and pathogens, as propolis can reduce the microbial load in a bee colony. The presence of a sufficient quantity of propolis in a bee's nest also seems to influence the activity of immune alleles: young bees activate their innate immune response less and can use the energy they save in this way for other functions, such as living longer. As a single sample of propolis can contain up to 300 chemical compounds, it is difficult for parasites and pathogens to develop resistance to such a combination of compounds.

The question is whether this can be called self-medication. The collection of certain types of nectar, pollen or resin may have been carried out independently of the direct presence of a pathogen (constitutive), or in direct response to an infection (inducible). If it is inducible, it can be considered a form of self-medication, defined as "defence against pathogens and parasites by one species using substances produced by another". There is evidence that bees do indeed choose selectively between different honeys depending on their state of health, and that different honeys contain substances that are active against different diseases. Honey bees also react to the presence of disease by allowing more bees to collect propolis. However, it is to be expected that the collection of nectar, pollen and propolis is mainly constitutive. Bees have to protect their hive all year round, even during periods when they can't collect anything. So, they have to make sure their medicine cabinet is full all year round. As they have no information about future infections, it is important that they stock up on honey, pollen and propolis from different plants.

Unfortunately, when beekeepers harvest these products and replace the honey with sugar, the bees' resistance also decreases. In addition, beekeepers have selected varieties that produce less propolis because they find it difficult to open the hives when there is a lot of propolis. This may save time, but it is to the detriment of the bees' resistance to disease and perhaps also to the detriment of protecting the nest against the Asian hornet.

Recommended reading

Bogdan I. Gherman, Andreas Denner, Otilia Bobiş, Daniel S. Dezmirean, Liviu A. Mărghitaş, Helge Schlüns, Robin F. A. Moritz and Silvio Erler, 2014. Pathogen-associated self-medication behavior in the honeybee *Apis mellifera*. Behavioural Ecology and Sociobiology 68: 1777-1784.

Marla Spivak, Michael Goblirsch and Michael Simone-Finstrom, 2019. Social-Medication in Bees: The Line Between Individual and Social Regulation: https://www.sciencedirect.com/science/article/pii/S2214574518300853

Silvio Erler and Robin F. A. Moritz, 2016. Pharmacophagy and pharmacophory: mechanisms of self-medication and disease prevention in the honeybee colony (*Apis mellifera*). Apidologie 47:389–411.

Chapter 22:
Black bees and racism

In 1910, an 11-year-old boy from the village of Mittelbiberach in Baden-Württemberg was entrusted by his mother, because of his poor health, to a Benedictine monk who was recruiting new members for his monastic community at Buckfast Abbey in England. The boy, called Karl Kehrle, was given the difficult task of cutting stones for the restoration of the abbey, despite his weak constitution. When it became clear that he was not up to the hard work, and was also plagued by homesickness, he was transferred to the monastery's apiary in 1915. There, as assistant to the monastery's beekeeper, he learned the trade of beekeeping. The reason I am reporting today what, a century later, would be considered parental failure and child labour, is that Karl Kehrle, whose monastic name was Brother Adam, became famous as a bee-breeder. There is no doubt that Brother Adam was a great bee-breeder. By chance, he came into possession of a publication by his compatriot, the zoologist Ludwig Armbruster. Armbruster had trained in genetics and had written a book on beekeeping. Brother Adam used this book as a guide to bee breeding. This is certainly not easy. To breed bees, you have to produce a large number of queens and have a lot of experience to know which are the best queens to rear. Then the colonies produced by the queens all have to be tested for the desired characteristics. And then there's the problem that queens need to mate preferably with drones with the same desired characteristics. Finally, you need thorough administration of the queens' pedigrees. It is a lot of work and technically difficult. Brother Adam worked long hours and did this monk's work with great dedication. He wanted to breed a super bee that would give the professional beekeeper maximum honey yield with as little work as possible. He strove to select bees with high fecundity, which collected nectar with great diligence, were resistant to disease, produced few swarms and were not aggressive. He actually developed such a super bee, which is still known as the Buckfast bee, a breed of bee that is still popular today. For this work, Brother Adam was awarded two honorary doctorates. We would have liked to end this short life story by saying that everyone lived happily ever after. This was certainly true of Brother Adam, who died at the blessed age of 98, and I suspect he looked back with satisfaction at what he had achieved in his life. However, the rest of this chapter is about the dark side of this story. Brother Adam was barely educated. He had taught himself the principles of genetics with the help of Armbruster's book, but he lived in the isolation of Buckfast Abbey. His beliefs were mainly based on his own practical experience and conversations with other beekeepers, but he lacked the relevant scientific knowledge.

Around 1860, beekeepers in Western Europe began importing bees from elsewhere. At the time, people didn't realise that importing exotic animals was dangerous, because of the diseases and parasites that could lodge in them. The danger became clear in 1907, when a previously unknown disease of bees was identified for the first time on the Isle of Wight, and so came to be known as Isle of Wight disease. In the years that followed, the disease spread throughout England. By 1915, the year Brother Adam began working in the Abbey apiary, 30 of the 46 colonies there had died from the new disease. All the local English colonies of black bees had died, while the surviving colonies were made up of lighter-coloured bees of the Italian subspecies *A. m. ligustica*. Many colonies of black bees also died in the vicinity of Buckfast Abbey. In retrospect, the most likely explanation is that the new disease arrived with the bees imported from Italy, so these bees already had some resistance to the disease.

Brother Adam wrote of the disease: "All the native bees in the area died in the winter of 1915-1916". This was a gross exaggeration because, as we now know, despite the high mortality rate from the new disease, black bees survived throughout England. With surviving colonies of the black bee, rapid selection for resistance in these bees would certainly have been possible. But Brother Adam concluded that the native black bees were totally inferior. For example, he wrote: "Before the extermination of the native race, all the diseases of the bee were present in our apiaries. With the extermination of the native variety, all diseases disappeared at once, with the exception of the tracheal mite". But this was pure nonsense.

Due to his lack of scientific training and his isolated life, he was unaware that his contemporaries, the scientists John Haldane, Ronald Fischer, Sewall Wright and Theodozius Dobzhansky, had developed a field that still forms the basis of our understanding of natural and artificial selection. Thanks to their research into the relationship between genetic variation and selection, the biologists knew that in large populations such as that of the honey bee, there is always sufficient genetic variation to allow the selection of characteristics such as those selected by brother Adam.

Undaunted by this knowledge, Brother Adam became convinced that genetic variation for these traits was insufficient in the black bees of Western Europe. He was convinced that a well-executed selection programme, based on crossing subspecies, was the only way to breed better bees. He wrote: "Breeding within a subspecies can only select characteristics that are already present in the genetic material. What is not there cannot be selected. Therefore, the existing genetic material of a breed determines the limits of any breeding effort. However, we know that each breed of honey bee has different desirable economic characteristics. To unite the different desirable economic characteristics of individual bee breeds, we need to crossbreed subspecies". With the following conclusion: "Cross-breeding is the only selection method that can lead to new economically valid combinations". He repeated this credo in all the lectures he gave and in everything he wrote. There are, of course, differences between subspecies of honey bees, but they all have the same genes and largely the same variants of these genes. The percentages of these variants vary from subspecies to subspecies, but through selection you can change the percentage of variants chosen. Cross-breeding between subspecies is not necessary, although it can sometimes speed things up.

According to Brother Adam, "the most plausible explanation for the Isle of Wight disease was that it had somehow been introduced into this country". So, he was well aware of the danger of importing bees. Yet he spent a lot of time collecting bees in countries around the Mediterranean. Brother Adam certainly enjoyed escaping from the abbey on a regular basis and travelling to foreign countries with pleasant climates. His travel accounts do not give the impression that he made an in-depth study of local bees. A short excursion to a number of apiaries and a fleeting impression were enough for him to decide whether or not the bees were suitable for his breeding programme. As he himself wrote: "During the first explorations, I was forced to base my judgements mainly on the external characteristics and behaviour of individual colonies". In the end, he did not use all these imports to select the Buckfast bee. According to his own notes, the Buckfast bee is the result of the hybridisation of the Italian subspecies *A. m. ligustica* and the English black bees that he hated so much.

The spread of hybrid bee breeds and the importation of exotic subspecies has meant that throughout Europe, Africa and the Middle East, local subspecies are threatened or have already disappeared through hybridisation with drones from production colonies. This hybridisation is also due to the fact that beekeepers with pedigree bees allow the drones produced by their colonies to fly freely. As a result, the source of biodiversity from which Brother Adam drew his super bees is now virtually lost. Subspecies of wild bees appeared between 0.7 and 1.3 million years ago and have evolved independently ever since. They are adapted to the local climate and flora. They are also adapted to cohabiting with other local pollinator species. The few people who defended the breeding of local bees were brushed aside by Brother Adam, who pointed out that they were using "long-standing arguments in favour of the native bee".

He was unfortunately very biased against the Western European black bee and wrote, for example: "Crosses between subspecies have always had a bad reputation for excessive stinging. This is well known. *Apis mellifera mellifera* is undoubtedly the culprit. It is extremely aggressive by nature and is still found throughout Europe". Note the "still" in this sentence, as if this evolutionary failure had to disappear as quickly as possible. Commenting on the French populations of this bee, he said: "The unquenchable stinging instinct of the French bee is notorious, as are its many other undesirable characteristics - at least undesirable from our point of view".

There are two European subspecies of bee that have found favour with Brother Adam. These are the Italian bee *A. m. ligustica* and *A. m. carnica* from Austria and Slovenia. From Brother Adam's descriptions of the breeding lines within these subspecies, it seems that he really knew that super bees can also be bred by selection within a subspecies. He mentions an undesirable characteristic of carnica, the strong tendency to swarm, but then adds: "However, it can be assumed that this can be reduced by selection". In fact, there are selections of carnica and ligustica that compare favourably with the Buckfast bee, including in their tendency to swarm. Was it to enable his trips to the Mediterranean that he proclaimed that only crosses of subspecies could lead to the desired results ?

The strange and terrible thing is that Brother Adam was perfectly aware that his way of working was threatening the biodiversity of the honeybee. At the end of his book Breeding the Honeybee, he writes: "The greatest danger threatening almost all breeds of bees today is the general and indiscriminate use of hybrids and the wide distribution of some very good strains [read: Buckfast]. This is leading to a loss of the genetic richness that was once available". Brother Adam continues: "In the years to come, it will be increasingly difficult to find true representatives of the different geographical breeds, which are essential for crossbreeding". His solution to the problem is: "In order to maintain and promote these breeding opportunities, it is essential to create reserves to preserve these different breeds". He then repeats the error on which all his work is based: "*By selecting within a subspecies, we can only intensify and record what is available. But by synthesising new combinations, the limits of selection with pure subspecies are lifted and we can obtain positive improvements in the honey bee*". Brother Adam was certainly not alone in crossing subspecies of honey bees, but his unprecedented popularity set the tone for 20th century beekeeping and gave the European black bee a bad name. There are still a large number of beekeepers who believe that Brother

Adam's credo is correct. In Brother Adam's day, there was no widespread concern about the decline in biodiversity, but now that this is the case, it is difficult to understand why beekeeping in Britain and elsewhere in Europe is still organised in the same way as it was in Brother Adam's day.

Recommended reading

Brother Adam, 1983. In search of the best strains of bees. Northern Bee Books, Hebden Bridge, UK.

Brother Adam, 1987 Breeding the honeybee. Northern Bee Books, Hebden Bridge, UK

Dobzhansky T, 1937 Genetics and the Origin of Species. Columbia University Press

Plutynski, A (2009). *The Modern Synthesis.* Routledge Encyclopedia of Philosophy. https://www.rep.routledge.com/articles/thematic/modern-synthesis-the/v-1.

Black bees on meadowsweet

Chapter 23:
Honey bees as competitors to solitary bees and bumblebees

The role of competition between individuals of the same species is an important element of Charles Darwin's theory of evolution by natural selection. Darwin mentions competition at least 49 times in The Origin of Species, whereas he mentions parasitism half as often and predation only twice. Mutualism is only mentioned in relation to the pollination of flowers. Twentieth-century biologists also attached great importance to competition and conducted a great deal of research into its importance for evolution. The idea, originally put forward by Thomas Robert Malthus, that populations continue to grow until resources become limiting, implies that competition between individuals of the same species is common. It was generally thought less obvious that there would also be strong competition between individuals of different species. After all, although we expect competition between species, if they use the same limiting resource, species differ in all sorts of characteristics that may reduce competition. For instance, they often use different parts of the area that they inhabit together, or they have different search strategies. Therefore, species only partially overlap in their use of a common resource. An important question then is how much overlap can there be, that makes coexistence of species in the same community possible, and when becomes the overlap too great so that species become mutually exclusive and can no longer live together/coexist. Mathematically trained ecologists have made many calculations regarding this problem and numerous studies have been published on the subject. These show that it is possible that one species displaces another through competition. Nevertheless, this does not contradict that species that use the same resources can often live together stably in nature. As mentioned above, this is partly because there are always differences between two species, making the overlap incomplete. Additionally, it is due to variability in nature, with circumstances sometimes more favourable for one species and sometimes for another. This complexity makes it extremely difficult for biologists to establish whether competition between species has led to the disappearance of the least competitive one. For instance, it is difficult to design experiments in a natural situation in such a way that they can demonstrate this unequivocally and can exclude other causes for the disappearance of a species from a community. In the early 1980s, there was a lively debate among American ecologists about the importance of competition between species. They organised a conference on this topic, of which the contributions were published in 1983 in a special issue of The American Naturalist. The question they tried to answer was not whether there is competition between species, because we know there is, but what the effect is of competition between species on the distribution, number and composition of species in nature. According to these and more recent publications, there is unfortunately still no unequivocal answer to this question.

Here, I will discuss this question for bee communities and potential competition with other insect species. Indigenous honey bees are unlikely to be significant competitors to bumblebees, solitary bees and other pollinators. The European subspecies of honey bee originated in the Pleistocene about a million years ago. They recolonised Europe again after the last ice age and, thus, have been part of the community of pollinators for at least 6 000 years. Over this period, the species have adapted to each other and to the plant species from which they collect pollen and nectar, allowing their long-term coexistence. Honey bees live on a very different scale of space and time as other pollinators and, thus, occupy a substantially different niche. These differences in niches that make coexistence possible are found in their habitat, in choice of food plant species and in the period of the season during which they are active and the temperature at which they fly. Thus, although honey bees, bumblebees and hundreds of species of solitary bees all feed on nectar and pollen, they have coexisted in the same community for thousands of years.

The most important difference between honey bees and solitary bees is that honey bees gather food for a colony of thousands of individuals. Therefore, they need large amounts of food and need to use sites with a high concentration of flowers. They feed on a spatial scale that can cover hundreds of square kilometres. The activities of honeybee colonies during the flowering season have two main objectives: reproduction and survival. So, food is needed for the production of swarms and drones and to stock up to see them through the winter. Bumblebees and solitary bees pass the winter in diapause and do not need food during hibernation. Solitary bees forage alone and can, therefore exploit small patches of flowers. They generally do this within a radius of a hundred metres of their nest, and live in an area of only a few hectares. Solitary bees are also often food specialists, active only during the flowering period of their host plants. In contrast, honey bees are not selective and exploit many different plant species as long as there are many flowers available. Bumblebees live on an intermediate spatial scale. The differences in spatial scale on which honey bees, bumblebees and solitary bees forage therefore limit competition for food.

Yet, articles regularly appear in the press warning that competition of honey bees poses a threat to bumblebees and solitary bee species. The research they refer to often shows that in places where many honey bees forage, there are often fewer solitary bees and bumblebees present than in similar places without honey bees. From such a negative correlation one cannot conclude that there is interspecific competition between honey bees and other bee species and that honey bees have a negative effect on other species, although these studies convincigly show that honey bees affect the behaviour of other bee species. For example, perhaps the other bees are avoiding competition in the area where there are many honey bees and forage elsewhere, perhaps the other bees have changed their activity pattern in response to the presence of honey bees and are foraging earlier, before the researchers made their observations and there are many other possible causes.

As mentioned before, studying competition is highly challenging. To answer the question if honey bees can outcompete other bee species, more rigorous research is needed.

First of all, it has to be demonstrated that resources are limiting, because otherwise interspecific competition cannot play a role. This evidence is lacking in much of the published research that claims a negative effect of honey bees. For instance, one such study was carried out in rape seed fields. It has been repeatedly shown that only about 18% of rape seed flowers are visited and pollinated by insects, whereas there is self-pollination in the vast majority of flowers. It is, therefore not plausible that food is limiting for pollinators in rapeseed fields and no proof that honey bees have a negative effect on other pollinators. Another example concerns apple farms. Apple farmers often use honey bees for pollination in orchards to improve fruit setting. Nowadays, an increasing number of apple farmers release also extra mason bees (solitary bees) of the genus *Osmia* in addition to the natural pollinators already present, as this further improves fruit setting. Therefore, it also does not seem likely that food is limiting in orchards where honey bees are present. The same holds for heathland. A healthy Calluna heathland produces 20.000 kg of honey and 2500 kg pollen per km^2. A honeybee colony needs 40 kg of pollen and 240 kg honey per year, hence a square km of Calluna heathland would be enough to support 62 colonies of honey bees, far more than the actual density, where apiculturists place hives. Hence, it is highlly unlikely that in flowering rapeseed fields, orchards and Calluna heathland food would be limiting for the pollinator community and, thus, unlikely that there is competition between pollinator species that would endanger a population.

Second, in situations where and when food is limiting all pollinator species would compete for food with each other. Thus, it would be quite challenging to study the effects of interspecific competition. Imagine a local community of 25 interacting species, potentially competing with each other. There are then 300 possible species pairs, and one should study all of them to find which one is outcompeting the other and not lump 24 species to try find the effect of the 25th, the honeybee. So far, no studies have been published that provide the necessary evidence needed to show that honey bees are superoir competitors in interactions with all other bee species in a community

Third, to show an effect of competition, data collected at one point in time are not sufficient. It should be shown that in the presence of a superior competitor the population density of the inferior species decreases. Therefore, one has to follow the competing populations over at least several seasons. Several studies that have done so and found a negative impact of honey bees on bumblebees. A good example is Diane Thomson's study . Niche overlap between Apis and Bombus varied substantially, but increased to levels as high as 80-90% during periods of resource scarcity. A significant negative relationship between honeybee and bumblebee numbers was observed in only one of the seven months, when resources were limiting and niche overlap large. In the years following the introduction of honey bees, the bumblebees produced fewer reproductive offspring, suggesting that bumblebee fitness decreases in the presence of honey bees. A second example that bumblebees have reduced fitness in the proximity of honey bee hives comes from Elbgami and collegues.

What can be learnt from these studies is that the presence of foraging honey bees affects the behaviour of bumblebees, such that the latter avoid areas with honey bees or stay shorter in them. When resources become scarce during part of the season, this behavioural interaction may result in a reduction of the fitness of the bumblebees. Honey bees might contribute to the decline of other species, and that should be taken as a serious warning.

In the following, I will address what has changed recently, after 6,000 years of stable coexistence between pollinators and may have resulted in the observed interspecific competition. In order of importance, these are :

The availability of flowers
Modern large-scale farming and the use of herbicides have led to the virtual disappearance of flowers from fields. The use of insecticides that kill bees in arable crops has made fields a dangerous environment for all types of bees. The fertilisation and drainage of pastures and hay meadows and the use of a

single type of grass (English ryegrass) have led to the disappearance of most flowers from agricultural grasslands. The development of agricultural land, in particular of large-scale monocultures, has led to the disappearance of much of our native flora and the destruction of small-scale, diverse landscapes. Large-scale, pesticide-intensive agriculture means that there is sometimes a shortage of pollinators for a crop at the time of flowering. Many beekeepers, therefore, believe that by keeping bees they are making a positive contribution to the environment, but this is only partly true. The lack of pollinators in agriculture is also due to the large-scale nature of the crop, which leads to major fluctuations in the supply of nectar and pollen over the season. This means that when the agricultural crop is flowering, there are too few pollinators, but once flowering is over, the lack of flowering herbs in the surrounding area creates a situation where there is no longer enough food for pollinators. The honey bee season starts in early spring. Flowering trees and shrubs then alternate, starting with willows and blackthorns, followed by plums, cherries and pears, then apples, hawthorns, horse chestnuts and maples, and finally lime trees, chestnuts, wild roses, brambles and locust trees imported from America. A similar series can be drawn up for flowering herbs, from crocuses, rapeseed and dandelions to meadowsweet. In spring, there is plenty of food in many places for insects feeding on nectar and pollen. Then, in the second half of the summer, a period of dearth begins as there are no longer flowering herbs in the fields and meadows. This shortage is man-made and a direct consequence of current farming methods. Competition between pollinator species is to be expected at times of shortage. When nectar and pollen reserves are low in early summer, honey bees can draw on their stocks, but when the weather is favourable, they also seek out the rare flowers on which bumblebees and solitary bees depend. It is striking that in an urban environment, where there is a wide variety of ornamental plants in bloom in summer, no negative effects of honey bees on other pollinators have been observed. Other exceptions, where food is probably not limiting for pollinators are heathland and salt marshes where sea lavender flowers in late summer, and ornamental plants such as Himalayan balsam (*Impatiens glandulifera*), but these habitats are rare.

The natural density of honeybee colonies
Wild honey bees generally nest in a hollow tree, at least one kilometre away from the next honey bee nest (Chapter 14). The density is therefore limited to 1 to 3 colonies per square kilometre.

Beekeepers place groups of hives in an apiary. There are often dozens of hives in the same place. This is not a problem as long as they are surrounded by large fields of flowering mustard or rapeseed, but in or near a nature reserve, this is asking for trouble. After the spring bloom, there are periods of food shortage and then, the bees from all these hives may compete with bumblebees, solitary bees, other pollinators and each other for scarce food sources.

The more similar the species, the greater the competition in the event of a shortage. This is why competition between conspecifics is always stronger than competition between individuals of different species. We can therefore expect honey bees to try to avoid competition with conspecifics from other colonies, for example by settling at a sufficient distance from other bee colonies. In this way, density is regulated to minimise competition with other bee species. The threat to bumblebees and solitary bees reported in the press is therefore not caused by honey bees, but by beekeepers who artificially increase colony density so that not only do honeybee colonies compete with each other, but other bee species are also affected.

The importation of exotic subspecies of honey bees with longer tongues than the black bee

One of the reasons why Brother Adam preferred Italian bees to English black bees was that Italian bees could harvest the nectar of red clover. Red clover is a native plant that is pollinated by bumblebees. The slightly longer tongue of the Italian bees makes them potential competitors for the long-tongued native bumblebees, as do bees of the other common subspecies, *carnica*, in contrast to the shorter-tongued native black bees.

The surge in honeybee viruses associated with varroa mites

Colonies of bees infected with varroa mites eventually die from infections by viruses carried by the mites. The workers in these colonies can transmit these viruses to other bee species when they visit the same flowers. Some of the observed negative effects of honey bees on bumblebees and solitary bees may well be the result of such viral infections.

All the four mentioned negative effects of beekeeping on wild bee species are, therefore the result of situations created by man. If we wanted to, we could modify agriculture and beekeeping such that they no longer pose a threat to other pollinating insects. This would also benefit agriculture, because it would solve the pollinator shortage!

Recommended reading

The American Naturalist 1983. A Round Table on Research in Ecology and Evolutionary Biology. The American Naturalist Vol. 122

Francisco Rubén Badenes-Pérez, 2020. Benefits of Insect Pollination in Brassicaceae: A Meta-Analysis of Self-Compatible and Self-Incompatible Crop Species. *Agriculture* **2022**, *12*: 446. doi.org/10.3390/agriculture12040446

Lindsey Button and Elizabeth Elle 2014. Wild bumble bees reduce pollination deficits in a crop mostly visited by managed honey bees. Agriculture, Ecosystems and Environment. http://dx.doi.org/10.1016/j.agee.2014.08.004

Lucas A. Garibaldi, Ingolf Steffan-Dewenter, Rachael Winfree, Marcelo A. Aizen, Riccardo Bommarco, Saul A. Cunningham, Claire Kremen, Luísa G. Carvalheiro, Lawrence D. Harder, Ohad Afik, Ignasi Bartomeus, Faye Benjamin, Virginie Boreux, Daniel Cariveau, Natacha P. Chacoff, Jan H. Dudenhöffer, Breno M. Freitas, Jaboury Ghazoul, Sarah Greenleaf, Juliana Hipólito, Andrea Holzschuh, Brad Howlett, Rufus Isaacs, Steven K. Javorek, Christina M. Kennedy, Kristin M. Krewenka, Smitha Krishnan, Yael Mandelik, Margaret M. Mayfield, Iris Motzke, Theodore Munyuli, Brian A. Nault, Mark Otieno, Jessica Petersen, Gideon Pisanty, Simon G. Potts, Romina Rader, Taylor H. Ricketts, Maj Rundlöf, Colleen L. Seymour, Christof Schüepp, Hajnalka Szentgyörgyi, Hisatomo Taki, Teja Tscharntke, Carlos H. Vergara, Blandina F. Viana, Thomas C. Wanger, Catrin Westphal, Neal Williams, Alexandra M. Klein, 2013. Wild Pollinators Enhance Fruit Set of Crops Regardless of Honey Bee Abundance. Science 339: 1608-1611.

Twfeik Elbgami, William E. Kunin, William O. H. Hughes, Jacobus C. Biesmeijer, 2014. The effect of proximity to a honeybee apiary on bumblebee colony fitness, development and performance. Apidologie (2014) 45:504–513

Rachel E. Mallinger, Hannah R. Gaines-Day, Claudio Gratton, 2017. Do managed bees have negative effects on wild bees?: A systematic review of the literature. PLOSone. | https://doi.org/10.1/journal.pone.0189268

Rachel E. Mallinger and Claudio Gratton, 2015. Species richness of wild bees, but not the use of managed honey bees, increases fruit set of a pollinator-dependent crop. Journal of Applied Ecology 2015: 323–330.

Alfredo Valido 1,2, María C. Rodríguez-Rodríguez1 & Pedro Jordano, 2019. Honey bees disrupt the structure and functionality of plant-pollinator networks. Scientific Reports | (2019) 9:4711 | https://doi.org/10.1038/s41598-019-41271-5

Diane M. Thomson, 2006. Detecting the effects of introduced species: a case study of competition between Apis and Bombus. OIKOS 114: 407 418,

Anika Hudewenz and Alexandra-Maria Klein, 2013. Competition between honey bees and wild bees and the role of nesting resources in a nature reserve. Journal of Insect Conservation 17:1275–1283

Arvid Bolin, Henrik G. Smith, Eric V. Lonsdorf and Ola Olsson, 2018 Scale-dependent foraging tradeoff allows competitive coexistence. Oikos 127: 1575–1585.

Sandra A. M. Lindström, Lina Herbertsson, Maj Rundlo, Riccardo Bommarco and Henrik G. Smith, 2016. Experimental evidence that honey bees depress wild insect densities in a flowering crop. Proc. R. Soc. B 283: 2016164. http://dx.doi.org/10.1098/rspb.2016.1641

Victoria A. Wojcik,1 Lora A. Morandin, Laurie Davies Adams, and Kelly E. Rourke, 2018. Floral Resource Competition Between Honey Bees and Wild Bees: Is There Clear Evidence and Can We Guide Management and Conservation? Environmental Entomology, 47: 822–833.

Different pollinators on the same blackberry bush (*Myathropa florea*, *Eristalis tenax*, *Bombus terrestris*, *Hylaeus* spec., and *Apis mellifera*.). They are potential comptitors. Only if food is lmiting they will actually compete.

Chapter 24:
Conservation sanctuaries

Brother Adam understood very well that the hybridization of subspecies of the honey bee that he was propagating, and the importation of exotic bees necessary for this purpose, led to a loss of biodiversity. His solution to the problem (Chapter 21) was to create reserves in which the subspecies would be protected. He did not specify what these reserves should look like. In the meantime, reserves for the native black bee exist in many places in Western Europe: in Belgium (3), France (13) and Switzerland (1). In the Netherlands, the island of Texel has banned bee imports since 1980.

How are these reserves created and are they a solution for the conservation of the black bee? The answer depends on the history of a reserve and the extent to which it is isolated from its environment. A number of French reserves are located on offshore islands where exotic bees have never been introduced, such as the island of Ouessant, the island of Oléron, the island of Groix and Belle Île. These populations are well protected and their survival is not threatened. However, there are also black bee sanctuaries on the mainland. These reserves are always surrounded by areas where exotic bees are kept.

Bénédicte Bertrand studied the extent to which foreign drones can enter an experimental black bee sanctuary in the Ile de France region (the Forêt Domaniale de Rambouillet). The core of this experimental reserve has a diameter of three kilometres, in which only black bees can be kept. Around this, there is a four-kilometre buffer zone, and around this, an eight-kilometre zone in which the type of bees (exotic, hybrid or black bees) is controlled. In the buffer zone, 70% of the drones came from the central zone and 90% came from the central zone and the buffer zone combined. As a result, 10% of the drones were animals that had entered from the surrounding area. The reserve had 74 colonies at the start of the study, 90 a year later and 143 in the final year of the study. Over the course of the study, the percentage of black bees remained constant. It is therefore possible to protect black bees in such a reserve, but only if there is a large buffer zone around it, in which colonies containing hybrids must be identified each year by DNA research. These colonies must then be replaced by colonies from the central zone of the reserve. The mating biology of honey bees allows queens to mate with drones born 15 kilometres away, as Annette Jensen has shown. It is therefore advisable to check regularly that there are no hybrids in the central area of the reserve. The size of a reserve depends on what we consider to be a vital population of black bees. Let's assume it's made up of three hundred colonies. Because we want to avoid a high density of honey bees to prevent them having a negative effect on other bee species (chapter 22), we opt for two colonies per square kilometre. We are therefore creating a reserve of 150 square kilometres. All the beekeepers located in such a reserve must participate and can only keep black bees, otherwise it won't work. Some of these beekeepers will be living in the buffer zone, which isn't interesting, because their bees could be hybrids, so they can't sell them until they've established whether that's the case or not.

One of the important conclusions of this book is that to give natural selection a chance to make honey bees resistant to disease and parasites, a population structure is needed in which honey bees can mate freely. Such a free-mating structure, in which queens can mate with males from a wide area, mating with an extremely high number of males and the extremely high recombination rate of honey bees, are the three tools with which honey bees enter the arms race against diseases and parasites. The creation of an isolated reserve of black bees, which in order to remain pure are not allowed to mate with colonies in surrounding areas, is not ideal in this respect. Land-based black bee sanctuaries therefore have a number of disadvantages : the work is ongoing, it's expensive and it's not the ideal solution for preventing future honey bee diseases.

Are there any alternatives ? The current situation is that beekeepers can be roughly divided into three religions : they are either fans of Buckfast bees or carnica bees, and the rest would prefer to keep black bees and try to practise sustainable beekeeping. This means that large numbers of drones from Buckfast and carnica colonies are released into the wild all over the country, even in nature reserves. As a result, the part of the bee population with a free mating structure consists mainly of hybrids between carnica, Buckfast and whatever other imported bees there are flying. These hybrids are not wanted by anyone, which is why the carnica believers and the Buckfast cult always replace the queens. These queens then have to mate at isolated mating stations, making the all-important free mating structure difficult. If we were to cultivate pure lines of black bees in the same way as buckfast and carnica bees, this would only further segregate the honey bee population, and even then we would not have a free mating structure. Breeding black bees in the same way as bees of other races is therefore not a good solution. We now know that the view that only hybrids between subspecies can give good results in honey bee breeding was a mistake by Brother Adam, and that the highly valued characteristics of buckfast and carnica bees can be selected just as well in black honey bee populations. This means that the arguments in favour of breeding exotic bees are no longer valid. Creating black bee sanctuaries outside the islands does not seem to be a good option. The only sustainable solution is to keep only bees of the indigenous subspecies. And it's possible, it would be good for the bees and the beekeepers.

Recommended reading

Benedicte Bertrand, Mohamed Albukari, Hèlène Legout, Sibyle Moulin, Florence Mougel and Lionel Garnery, 2015. MtDNA COI-COII marker and drone congregation area: An efficient method to establish and monitor honeybee (*Apis mellifera* L.) conservation centres.

Fabrice Requier, Lionel Garnery, Patrick L.Kohl, Henry K. Njovu1, Christian W.W. Pirk, Robin M. Crewe and Ingolf Steffan-Dewenter, 2015. The Conservation of Native Honey Bees Is Crucial. Trends in Ecology and Evolution 34: 789-798. https://doi.org/10/1016/j.tree.2019.04.008

Robin F. A. Moritz1, F. Bernhard Kraus, Per Kryger and Robin M. Crewe, 2007. The size of wild honeybee populations (*Apis mellifera*) and its implications for the conservation of honey bees. Journal of Insect Conservation 11:391-397.

Chapter 25:
Darwinian beekeeping

The European Union is aiming to implement new rules for farm animal husbandry by 2027. These rules state that animals must no longer be adapted to the system in which they are kept, but that the system must be adapted to the animal's natural behaviour. Although honey bees are not mentioned, it would be worth examining whether beekeeping actually meets these requirements. In addition, current knowledge of bee behaviour and genetics means that we can make suggestions as to how beekeeping could become more sustainable.

Thomas Seeley has drawn up a list of 21 points of difference between the lifestyle of wild bees and that of honey bees. Keeping bees as they live in the wild would improve their well-being and health. Seeley calls this way of keeping bees "Darwinian beekeeping".

Resilient bees

In the space of 40 years, two major exotic natural enemies of bees have colonised our country: the varroa mite and the Asian hornet. A third species, the small honeycomb beetle, has already been spotted in Italy and is expected to spread throughout Europe. Finally, a number of species of parasitic mites live in southern Asia and could also be transmitted to European bees and spread. In addition, viruses already present could evolve into new, more virulent variants, as could pathogenic bacteria, fungi and microsporidia. For example, the deformed wing virus present in the UK in 2006 consisted mainly of variant A, whereas in 2017 it consisted mainly of variant B. This varroasis-associated virus is one of the main causes of bee mortality.

This means that new diseases and parasites can and will appear in the future. So, honey bees need to be able to adapt to a changing world.

Having learned from the varroa mite that the evolution of resistance to new parasites through natural selection is hampered by the way bees are kept in Western Europe and the United States, it is time to think about a beekeeping method that uses the potential of natural selection to make bees resistant to new parasites, predators and diseases.

In nature, queens selectively mate with males from a large area, they mate with a large number of males and they mix their genes during egg production in a way that is so radical it has not been observed anywhere else in the animal kingdom (chapters 6 and 9). This behaviour is the result of natural selection and must therefore have an important function. The best explanation is that it allows honey bees to keep up with the arms race with rapidly mutating pathogens, and to develop resistance to new parasites and pathogens through natural selection, just as honey bees in South Africa and South America rapidly became resistant to the varroa mite (Chapters 17 and 18). We saw earlier that sexual selection due to the enormous competition between drones (chapter 7) is an important consequence of the sex ratio in honey bees, with few queens but a large number of drones. Drone selection also plays an important role in selection for resistance, as drones infected with a virus or weakened by varroa mites are much less likely to mate. As drones have only one set of chromosomes, genetic weaknesses are always expressed and these weak alleles can be eliminated by selection. So, if disease prevention is an important objective, we should allow the bees to mate freely, integrating them into a large population in which natural selection can operate. Nor should we hinder colony reproduction by removing queen cells and drone brood. We should strive to keep the variation in heritable traits within the bee population as wide as possible, and therefore not breed large numbers of queens from a single colony. We should also stop importing bees from the USA or other distant countries and not replace queens with ones from elsewhere. We should let the bees decide for themselves how much of the brood is drone brood and therefore stop restricting drone brood.

The problem is that this disease mitigation solution is incompatible with all the other important objectives of professional beekeepers, namely maximising the honey harvest and minimising the labour and equipment costs required to achieve it. This can only be achieved with large colonies of bees selected for their low aggression and high honey production. If you leave these selected bees to mate freely, the desired characteristics will have disappeared after just a few generations.

Subsequent generations of honey bees would be made up of hybrids, which are often aggressive and give much lower yields. Consequently, pure-bred bees can only remain pure if the queens are mated in isolated mating stations or if they are fertilised by artificial insemination.

We therefore have to choose between resilient bees that develop resistance to new diseases and parasites through natural and sexual selection, and improved bees that are pure-bred and selected for high yield. Professional beekeepers will often choose pure-bred bees. But the number of professional beekeepers is small compared with the army of hobby beekeepers who do not depend on pure-bred super-bees for their income. Hobby beekeepers could therefore choose to let their bees mate freely, integrating them into a vast population in which natural selection can take place. Many hobby beekeepers already do this, but it would require professional beekeepers who raise pedigree bees and import queens to stop releasing their drones into the population. I think this is possible, for example by installing drone screens.

The swarming of bees

On an introductory beekeeping course, you'll learn all sorts of things to maximise the honey harvest. You'll learn how to break the cappings with young queens in spring, how to merge hives to make them bigger or how to divide them to prevent swarming. You may learn that you need to mark the queen and cut off one of her wings so that she cannot leave with a swarm. You may learn that by adding an extra brood chamber, you can reduce the risk of swarming, and that you can use drone frames, then later destroy these frames full of drone brood to curb the proliferation of varroa mites. All these actions limit the reproduction of honey bees. These manipulations are intelligently designed to enable the beekeeper to have large colonies, capable of producing a lot of honey. It's strange, moreover, that a beekeeping course should place so much emphasis on learning techniques designed to prevent the reproduction of honey bees. Natural selection seeks to maximise reproduction, not the honey harvest.

We saw earlier that the queen stops laying eggs, after she has laid eggs in the cells where the young queens grow (Chapter 12). The workers feed her less and chase her around, so that she has lost a quarter of her weight by the time she leaves the nest with a swarm. Once the swarm has arrived in its new home, it takes several days for new combs to be built. As a result, there is a period of around three weeks during which there is no brood. A similar pause in reproduction also occurs in the nest that the swarm has left. Firstly, because the outgoing queen stops laying eggs, and secondly, because the young queen who takes over the nest has to make a nuptial flight before starting to lay eggs. During the period when there is no brood, the diseases and parasites that live in the brood cannot multiply. Swarming honey bees therefore also prevents the development of diseases and parasites. What's more, a swarm generally moves to a new home that is still free of parasites and pathogens. Preventing a colony from swarming cancels out these advantages.

Personally, I always enjoy watching bees swarm. But you have to be there when it happens. As a swarm generally settles in a tree close to the hive after leaving it, there is little risk of losing it. There is usually enough time to shake the swarm into a bucket and place it in a prepared hive. The wonderful thing is that this interrupts the process by which the bees themselves seek a new home (Chapter 12) and yet this cruel disturbance usually leads to acceptance of the offered hive.

Type and size of hive

We have seen that the honey bees of Western Europe are originally forest dwellers (Chapter 1) that make their nests in tree cavities. If honey bees could choose their nest site, they would prefer not to live in a hive that is too large and poorly insulated, but rather in a tree cavity of around 40 litres (Chapter 13). If bees were housed in well-insulated hives of up to 50 litres, and if the bees were allowed to determine the size of the colony themselves, beekeepers' hives would on average be much smaller than they are today. This means that more hives would be needed for the same honey harvest, which doesn't seem to be a problem for the amateur beekeeper. When honey bees are used to pollinate orchards and other agricultural crops, having smaller colonies can actually be an advantage, because with more but smaller colonies, a better spatial distribution of pollinators is possible.

Honey bee nests in a tree cavity lose four to seven times less heat than nests in a thin-walled hive. Well-insulated hives can therefore significantly reduce the cost to the bees of keeping the colony warm over winter. These hives can also reduce the risk of overheating in summer. There are good plastic hives available today which, although not very attractive, are effective. Hobby beekeepers can also build their own wooden hives insulated on the inside with the panels used to insulate cavity walls. The rough interior of these panels provides a good substrate for the bees to deposit propolis, which is also beneficial for the health of the hive.

Apiaries or isolated hives.

Colonies of honey bees live naturally far apart in the landscape. This reduces competition with other bees and the risk of parasites and disease. Grouping bee colonies together in an apiary only serves the convenience of the beekeeper, who thus creates all sorts of risks. The bees make mistakes and enter other hives, from where they can then bring back varroa mites or pathogens, or they plunder honey from weakened hives and return to the nest with diseases and parasites. On average, if there are a lot of hives in an apiary, the bees also have to fly further to collect honey and pollen. So it's always best to place the hives as far apart as possible.

Varroa resistance

If all European beekeepers stopped using chemical methods to combat varroa mites and let the bees reproduce and mate freely, natural selection could do its work. Colonies with a greater number of resistance alleles will survive better and eventually become resistant. Unfortunately, the frequency of occurrence of these resistance alleles is currently very low in pedegree bees, the result of the selection of exotic bees for high productivity, limited swarming behaviour and low aggressiveness (chapter 9 and 17). Stopping chemical control would therefore result in the collapse of the European pedegree bee population, with only a small number of hives surviving. To remedy this situation, it is therefore advisable to increase the frequency of resistance alleles in these bees through artificial selection (Chapter 19).

Competition with othe pollinators

The natural density of wild honey bees in Western Europe ranges from 0.5 to around 3 colonies per square kilometre. At these densities, there is no negative impact of honey bees on other bee species to be expected. Regulations on the number of bee colonies that can be established in or near nature reserves could be based on observed natural densities. Given that the pollinator community is adapted to coexistence with the black bee, it is recommended that only this subspecies be conserved in Western Europe. Thanks to their slightly shorter tongue, they compete less with certain bumblebee species (Chapter 24).

Beekeepers can rear a large number of queens, far more than the bee colonies themselves. What's more, they can feed the hives with protein-rich pollen substitutes and sugar solutions, making it possible to produce many more colonies than the carrying capacity of the environment can support. Unfortunately, modern large-scale farming has only reduced this carrying capacity. If additional colonies were obtained simply by dividing up nests and trapping swarms, not only would natural selection work better, it would also prevent overpopulation of honey bees and hence competition with other pollinators.

A large part of the UK is agricultural. Many agricultural and horticultural crops depend partly or entirely on insects for pollination. Because of the layout of our farmland, it is hard for pollinators to find resources outside the period when a crop is flowering and needs pollinating. This situation threatens agricultural areas with a shortage of pollinators when they are needed, and competition between pollinators in times of shortage. Farmers, beekeepers and biodiversity all stand to gain if our farmland is managed in such a way that there is nectar and pollen to harvest from March to October, and if these sources of nectar and pollen are distributed across small-scale landscape features.

The black bee

Brother Adam proclaimed the fallacy that genetic variation within populations of black bees and other subspecies was too low to select colonies that produced a lot of honey, were low in aggression and had other characteristics that he found attractive (Chapter 22). He repeatedly and vigorously asserted that hybridisation between different subspecies was the only way to breed bees successfully. His assertions had no scientific basis. The knowledge already available at the time and the extensive research carried out since on genetic variation within populations and on natural selection confirm that he was wrong. If Brother Adam had devoted his energies to creating breeding lines of European black bees, they would have become at least as productive as his buckfast hybrids. What's more, they would have had the advantage of being better adapted to the local flora, the local pollinator community and the local climate.

Now that we know this, there are few reasons left to keep exotic and hybrid bees based on these misconceptions.

Nevertheless, beekeepers today mainly keep the exotic carnica subspecies from Central and Eastern Europe and hybrid Buckfast bees. Isolated mating stations, where young virgin queens can only mate with males of their own race or subspecies, are widely used to maintain the purity of these bees. Since a limited number of queens and a limited number of drone colonies are involved, and since selection in the past was based mainly on characteristics other than resistance to disease and parasites, it is likely that the mating possibilities offered by this breeding method do not provide sufficient genetic variation. The high vulnerability to varroa mites and associated viruses is proof of this. Beekeepers allow their queens to mate in isolated mating stations, but allow the drones in their colonies to fly freely without concern for the native bees. For natural selection to produce resistant bees, you need a large population of bees that mate freely. Given that different breeds of bee are bred today and that beekeepers release their drones into the wild, this freely mating population is necessarily made up of hybrids. Nobody likes that. That is why we should launch a breeding programme in Western Europe, to cultivate high-yielding black bee lines for professional beekeepers.

If hobby beekeepers only keep black bees, we'll have a population of healthy, free-mating bees in which natural selection can do its work. What's more, the black bees would return to the forests to live in the wild in hollow trees. If professional beekeepers used highly productive breeding lines of the black bee, they would also have fewer problems letting their drones fly, because the risk of hybridisation would disappear and natural selection would correct genetic traits that do not work as well in the wild.

In fact, there is no good alternative. As well as breeders of Buckfast and carnica bees, there are now beekeepers who want to breed black bees. However, if black bee breeders also became dependent on isolated mating stations, we would lose the most important tool for ensuring long-lasting, hardy bees: natural selection.

Recommended reading

Thomas D. Seeley. The lives of bees. The untold story of the honey Bee in the Wild. Harvard University Press, Cambridge Massachusetts.

Chapter 26:
Finally

The three main messages of this book are worth highlighting here.
Resistance to new pathogens

Honey bees are highly susceptible to infectious diseases because of the way they live: frequent direct contact with fellow bees, the passage of food between individuals, and life in a nest at a constant high temperature, which is optimal for the multiplication of pathogens. They have three lines of defence, two of which are at colony level: hygienic behaviour and the collection of antibiotics in the food and resin with which they surround their nests. The third line of defence is the immune response of individual bees. As pathogens have very short generation times and very large populations, they can evolve rapidly and develop new virulent variants. Honey bees are therefore engaged in an evolutionary arms race with their pathogens. They compensate for their long generation time (one year) and much smaller populations by a reproductive strategy that maximises genetic variation in the immune response. Honey bees achieve this in three ways: young queens mate at great distances from the nest with drones from a very wide area. The young queens mate with a very large number of males (10 to 20) and store the sperm of all these males for later use. The final and perhaps most important method is recombination. When queens make eggs in a process in which a cell with two sets of chromosomes divides into two cells each with one set of chromosomes, the chromosomes break at predetermined points and the pieces are exchanged with corresponding pieces of the other chromosome. This process of chromosomal mixing occurs in all animals. In honey bees, the frequency of recombination is 20 times higher than in humans. To get an idea of how extreme this is, imagine you're on a course of antibiotics, taking one tablet three times a day, and your GP calls to tell you that you have to take one tablet 60 times a day from now on. Population structure, polyandry and extreme recombination combine to form honey bees' weapons in the arms race against pathogens: they maximise heritable variation. Honey bees can only sustain this race if they are part of a large, freely mating population.

Beekeepers, in their quest for highly productive bees, have stripped honey bees of the aforementioned weapons. They prevent reproduction as much as possible, allow their bees to mate in limited, isolated populations and replace queens. Recombination is only effective if different gene variants are exchanged.

If both chromosomes have the same variant, the exchange makes no sense. Consequently, even a modest degree of inbreeding can make the weapon of recombination less effective.

It is remarkable that the regions of the world that use selected bees, mating stations and queens replaced every year are those where the bees are susceptible to varroa and its associated pathogens due to the absence of resistance genes to the mite and deformed wing virus. Genetic variation is lost when beekeepers select highly productive bees. Furthermore, to maintain pure lines of selection, it is necessary for these bees to mate in relatively small, closed populations called mating stations. I believe that only honey bees that can mate freely in a large, well-mixed population can resist new pathogens or new variants of pathogens that are already present.

Exotic honey bees
Most beekeepers are convinced that Buckfast bees and bees of the carnica subspecies are superior to the native *A. mellifera mellifera*, the black bee. This misunderstanding was created by Brother Adam who, in his writings, uses the word "strain" for both the subspecies and the breeding line. Buckfast bees are hybrid breeding lines between the Italian subspecies *A. m. ligustica* and the Western European subspecies *A. m. mellifera*. The carnica bees bred by German and Dutch beekeepers are selection lines of the Eastern European subspecies *A. m. carnica*. If we want to make a comparison, we need to compare selection lines such as Buckfast with selection lines of *A. m. carnica* and *A. m. mellifera*. However, the latter hardly exist, and this is also the fault of brother Adam, who hated this subspecies and wrote a lot of nasty things about it. In particular, he claimed that within the Western European black bee population *A. mellifera mellifera*, there would not be enough genetic variation to select highly productive lines, which is simply not true.

The great popularity of Buckfast bees can also be attributed to a misunderstanding created by brother Adam. Without any scientific proof to back up his claims, he stated that bee breeding was only possible through the hybridisation of subspecies. He repeated this message ad nauseam, despite all the evidence to the contrary. Carnica bee producers prove him wrong every day, but many buckfast enthusiasts still believe Brother Adam's propaganda. So you can develop equivalent high-yielding, low-aggression breeding lines from any European subspecies of *A. mellifera* (or its hybrids). If it doesn't matter, why is it preferable to keep only the native *A. m. mellifera* in Western Europe?

There are a number of arguments in favour of this :

Maintaining different subspecies and breeding lines requires isolated mating stations. This prevents bees from mating freely, which is not good for their resistance to new pathogens (see above).

Holders of exotic bees allow their drones to fly freely, thus eradicating the indigenous subspecies.

Importing bees from abroad disrupts the natural selection and local adaptation of bees. Importing bees from elsewhere also increases the risk of new diseases and parasites that may accompany the imported bees. This is why it would be preferable to stop importing bees from other countries.

Many beekeepers who keep exotic bees often replace their queens. In doing so, they go against natural selection.

The young queens of exotic bee breeds mate at mating stations, some of which are located in nature reserves. Nature reserves are there to preserve biodiversity, not to breed exotic species that threaten native honey bees.

Farmed exotic bees have longer tongues than native bees. This enables them to compete with bumblebees for nectar in flowers typical of bumblebees, such as red clover. These problems will disappear if we gradually allow the native black bee to return. For the moment, this is not entirely possible due to the lack of highly productive, varroa-resistant black bee breeding lines. These bees are needed to maintain the income of professional beekeepers. Together, the many hobby beekeepers can provide a natural and open population structure, within which natural selection can do its work and thus ensure a healthy population of black bees. They should then let their bees mate naturally and, if possible, swarm.

Competition
Several studies on possible competition between honey bees, bumblebees and solitary bees show that other bee species avoid areas where many honey bees forage. Although this is not irrefutable proof that honey bees are damaging the biodiversity of other bee species, these observations do not rule out the possibility that honey bees are having a negative effect on pollinator biodiversity. The problem in Western Europe is that there are few natural areas and large parts of the landscape are used for intensive agriculture. As a result of large-scale farming, in which many pesticides are used, there is sometimes a shortage of pollinators during the flowering period of an agricultural crop. Many beekeepers therefore believe that by keeping bees they are making a positive contribution to the environment. The shortage of pollinators in agriculture is also due to it's large-scale nature, which causes major fluctuations in the supply of nectar

and pollen over time. As a result, when the agricultural crop is flowering, there are too few pollinators, but when the flowering is over, the lack of flowering herbs in the vicinity creates a situation where there is no longer enough food for pollinators. In such a situation, honey bees can then have a negative effect on other pollinators through competition for limited food resources. Estimates of the density of wild bees in Europe vary from 0.5 to 3 colonies per square kilometre (Chapter 13). At these densities, no negative effects of honey bees on other pollinators have been observed. The figures for wild bee densities come from more or less natural environments. Densities should be much lower in intensively farmed areas if wild bees were present. As beekeepers can rear as many queens as they wish, they are able to create more colonies than the carrying capacity of the environment allows. Beekeepers must take care to maintain the number of honeybee colonies within natural densities. Southern European bee subspecies have slightly longer tongues than the native black bee. As a result, the overlap in food choices between imported honey bees and long-tongued bumble bee species is greater, and competition between these honey bees and bumble bees can occur. Regulating the local density of honeybee colonies and no longer breeding exotic honey bees are solutions that beekeepers can propose to resolve the problem of competition. However, the main cause of the problem lies in agriculture and can only be resolved by policy changes at national and European level.

Concluding remarks
By 1838, Charles Darwin had already developed his ideas on evolution by natural selection. However, it was not until 1859 that he dared to publish these ideas in The Origin of Species. One of the reasons he waited 21 years was that he didn't know how to incorporate bees into his theory. The problem was that the queen and workers differed considerably in the shape of their bodies, whereas they did not differ in genetics. (It was already known that drones developed from unfertilised eggs). He thought that the natural selection of workers could not be the explanation, because workers did not reproduce. It was therefore difficult to explain how the instinct to build combs with hexagonal wax cells had arisen through natural selection, as queens are not involved in building combs and have no wax glands. Darwin conducted experiments by giving bees coloured pieces of wax and observing how they used them to build comb. He thus became the first behavioural biologist. He had difficulty solving other problems, because it wasn't known in detail how genetics worked. It was completely unknown that gene reading depended in part on the environment, so that depending on the food

and the size of the cells, the same egg could turn into a queen or a worker. There was also the problem of how natural selection worked. Selection on individual workers could not be natural selection, because these workers had no offspring. Darwin came close to the idea of group selection (i.e. selection on entire colonies of bees), but he didn't dare.

Since Darwin, our knowledge of bees, genetics and natural selection has progressed enormously. Intelligent and creative researchers have devised experiments that have enabled them to obtain answers to their questions from the bees themselves. To do this, they had to observe the bees. This is how generations of brilliant researchers have been able to unravel the workings of bee colonies. In this book, we meet these researchers and discover the results of their research. In the meantime, knowledge of genetics has also progressed enormously. Thanks to the ingenious technologies of molecular biologists, we can now read the information hidden in the DNA strings of the bee genome. As a result, we now know a great deal about the evolutionary history of honey bees. This book also reports on this. We know that black bees are native to Western Europe and that they have been a distinct branch of the honeybee evolutionary family tree about a million years. We also understand how honey bees defend themselves against new diseases.

Much of this knowledge has not yet been transposed to the way bees are kept in Western Europe. We hope that this book will contribute to this transposition, so that reared honey bees are better able to display their natural behaviour and so that reared honey bees are no longer a threat to biodiversity.

Recommended reading

Charles Darwin, 1859. On the Origin of Species by Means of Natural Selection. John Murray, London.

Index

A

Adaptation 17
Aggregation pheromones 28, 143
A.m. carnica 6, 7,
American foulbrood 78, 97
A.m. ligustica 6, 7
Antibacterial 97
Antibiotic 34, 39, 65, 131
Antifungal 97
Antimicrobial 96
Apis armbrusteri 9
Apis cerana 77, 93
Apis mellifera mellifera 45, 104
Aristotle 19, 51
Arms race 85, 120, 124, 130, 131
Arnot Forest 64, 67, 68, 69
Asian hornet v, 93, 94, 95, 98, 123
Asian Hornet 93
a-virulence 41

B

Badger 65
Bear 65
Bee bread 97,
Beowulf Cooper 27,
Black honey bee 2, 121
Black locust 52, 96
Boerhave 19
British Beekeeepers Association 1
Brother Adam 74, 101, 102, 103, 104, 105, 112, 119, 120, 128, 132
Buckfast 74, 101, 102, 103, 104, 120, 128, 129, 132
Bumblebees 13, 108

C

Cape honey bee 20, 22, 81
Captive 73, 82
Caroline Otero 33
Charles Darwin 47, 49, 107, 134, 135
Chromosome 25, 43, 44
Coexistence of species 107
Colony 88
Competition 111, 114, 127, 133,
Conservation sanctuaries 119
Corona pandemic 34
Cross-breeding 103
Crossing subspecies
Cyprien Zmarlicki 27
Cypriot honey bee 94

D

Darwinian beekeeping v, 123
David Peck 68
Division of labour 14, 15, 16, 17, 20
Domesticated 73, 74
Dorothy Galton 67,
Drone 31, 89, 124,
Drone congregation area 48, 120

E

Elbgami 110, 113
Endophallus 28, 29, 88, 142
Eugène Marais 15
European foulbrood 97
European hornet 93
Eva Crane 75, 77
Exotic natural enemies 123
Exotic subspecies of honey bees 112
Extreme polyandry 35, 44

F

Fallopian tube 89, 142
Fanny Mondet 78

Fernando Pessoa 73
fight against to parasites and pathogens 43
Frequency of recombination 39, 41, 43, 44, 83, 84, 131
Friedrich Ruttner 28, 31

G
Gerald Loper 28
Giant hornet 92, 93
Gilbert White 27, 28, 30
Glaciation 5
Greg Hunt 79

H
Hans Ruttner 31
Heathland 109
Heterozygous 45, 142, 143
Homozygosity 44, 143
Homozygous 44, 143
Honeybee viruses 112
horizontal transmission 41, 143
Hybrids 74, 75, 87, 104, 119, 120, 124, 128, 132
Hygienic behaviour 21, 34, 44, 69, 78, 80, 81, 82, 131

I
Immune response 34, 40, 43, 98, 131
Immunity 34
Ingemar Fries 83, 84, 85
Interglacial 5
Internal clock 54
Isle of Wight disease 102, 103

J
Jan Swammerdam 19, 21
Jean Louveaux 12
John Haldane 102
John Harbo 78, 79, 89, 90
Jürgen Tautz 60

K
Karine Monceau 94, 95,
Karl Kehrle 101
Karl von Frisch 51, 57
Kirk Visscher 58, 59

L
Language of bees 51, 52, 55
Leo Tolstoy 67
Lime 77, 111
Ludwig Armbruster 101

M
Madeleine Beekman 22, 60, 61
Marla Spivak 78, 99
Martin Lindauer 55, 56, 57, 61, 63
Mating behaviour 28, 41, 74
Mating flight 26, 27, 28, 33, 34, 89
Maurice Maeterlinck 15, 17
Mike Allsopp 81
Mutation 43, 143
Mutualism 107,

N
Native 12, 45, 94, 121
Natural selection 48, 80, 125
Nectar 97
Niche 110, 143
Niche overlap 110, 143
Non-captive 73

O
Oriental honey bee 34, 77, 78, 79, 93, 94
Oriental honeybee 79
Oriental hornet 94
O.W. Park 78

P

Paleogene 144
Panmictic population structure 140
Parasitism 107
Parasitoid wasps 1
Pathogen 99
Periods of glaciation 5
Pheromones 20, 28, 29, 144
Pleistocene 108
Pollen 11, 12, 16, 17, 33, 39, 52, 54, 57, 97, 98, 108, 109, 111, 126, 127, 134
Pollinator 3, 7, 73, 74, 88, 97, 103, 108, 109, 110
Polyandry 34, 41, 43, 44, 131, 85, 143
Polygynous 37, 144
Primorsky 77
Propolis 39, 64, 65, 94, 97, 98, 126

Q

Queen 11, 13, 15, 16, 19, 20, 21, 23, 25, 26, 27, 28, 29, 30, 33, 34, 35, 37, 38, 39, 40, 41, 42, 44, 48, 60, 68, 82, 83, 84, 85, 87, 89, 90, 94, 101, 120, 123, 124, 125, 127, 128, 131, 132, 133, 134, 135

R

Ralph Buechler 40
Recessive 44, 144
Recombination 43, 44, 45, 46, 83, 120, 131, 132, 143
Reserves 11, 12, 104, 111, 119, 120, 127, 133
Resistant 2, 3, 44, 45, 48, 68, 69, 79, 80, 81, 82, 83, 84, 85, 88, 90, 91, 98, 102, 123, 124, 127, 128, 131, 132, 133
Robert Currie 34
Robert Page 15, 48
Roger Morse 27, 63
Ronald Fischer 102

S

Salt marshes 12, 111,
Scouts 17, 51, 52, 53, 57, 58, 59, 60, 61
Secondary plant substances 97
Sewall Wright 102

Sex determination 25, 26, 144
Sex ratio 26, 47, 49, 124
Sexual selection 47, 48, 49, 124
Single drone insemination 91
Sladen 28
Small honeycomb beetle 123,
Solitary bees 3, 11, 12, 33, 73, 108, 109, 111, 112, 133
Species 47, 105, 107, 113, 134, 135
Suresh Desai 34
Stephen Fleming 28
Stingless bee 1, 37, 87
Sub-populations 74
Subspecies 74, 81, 87, 102, 103, 104, 108, 112, 119, 120, 121, 127, 128, 132 133, 134, 144
Swarm 11, 20, 27, 34, 37, 38, 39, 40, 44, 48, 57, 58, 59, 60, 61, 63, 64, 65, 68, 69, 73, 74, 82, 84, 87, 104, 108, 125, 133,

T

The Asian hornet 93, 94, 95, 98, 123,
Themudo 45
Theodosius Dobzhansky 87
Thomas Robert Malthus 107
Thomas Seeley 2, 20, 29, 34, 42, 55, 58, 59, 60, 63, 64, 67, 68, 123
Tom Rinderer 64
Tremble dance 55

V

Varroa v, 66, 77, 78, 80, 85, 87, 127
Vertical transmission 143
Vespa crabo 93
Vespa mandarina 93
Vespa orientalis 94
Vespa velutina 93, 95
Virulence 41
Virulent pathogens 40, 41
Vladiwostok

W
Waggle dance 52, 53, 54, 55, 57,
Warwick Estevam Kerr 87, 88
Wild bee species 73, 112
Wolfgang Kirchner 55

Glossary

Aggregation pheromones	Volatile chemicals emitted by individuals that attract other individuals and is instrumental to keep them together in an area.
Endophallus	male sexual organ in insects
Polyandry	breeding system where the female mates with a large number of males
Fallopian tube	part of a female sex organs. It channels egg and sperm transport
Recombination frequency	the percent of meiosis in which homologous recombination exchanges segments between paired chromosomes
Heterozygous	Heterozygous having two different alleles of the same gene, one from the father, one from the mother
Homozygosity	possessing two identical alleles of a particular gene, one inherited from each parent.
Homozygous	having two the same alleles of a gene one inherited from each parent
horizontal transmission	transmission between members of the same generation
Mutation	an alteration in the nucleic acid sequence of a gene
Mutualism	a relationship where species involved benefit from their interactions
Niche	a species' unique ecological role determined by the way it interacts with abiotic and biotic resources in its habitat to survive and reproduce
Niche overlap	the partial or complete sharing of resources or other ecological factors by two or more species

Paleogene	a geologic period and system that spans 43 million years from the end of the Cretaceous Period 66 million years ago (Mya) to the beginning of the Neogene Period 23.03 Mya
Pheromones	chemicals emitted by individuals that convey information other individuals
Polyandry	when a female mates with two or more males
Polygynous	social insects with more than one queen in the colony
Recessive	a gene that is only expressed when an individual has two identical alleles of that gene, one inherited from each parent
Recombination	an exchange of segments between paired chromosomes creating new genetic variants
Sex determination	the genetic system that determines the sex of an individual
Sex ratio	Ratio between the number of males and the number of females in a population, expressed as % males
Subspecies	a genetically distinct population and geographically isolated population of a species that evolves independently
Vertical transmission	the transfer of a pathogen from the parent directly to the offspring

www.ingramcontent.com/pod-product-compliance
Lightning Source LLC
Chambersburg PA
CBHW042110230426
43662CB00043B/2465